ENVIRONMENTAL HEALTH - PHYSICAL, CHEMICAL AND BIOLOGICAL FACTORS

# TERRESTRIAL BIOMES

## GEOGRAPHIC DISTRIBUTION, BIODIVERSITY AND ENVIRONMENTAL THREATS

**Library of Congress Cataloging-in-Publication Data**

ISBN: 978-1-63484-625-7

*Published by Nova Science Publishers, Inc. † New York*

# CONTENTS

# PREFACE

This book provides current research on terrestrial biomes. Chapter One demonstrates the severe conditions of arctic areas that lead to the formation of common characteristics for all complexes of soil microfungi. Chapter Two discusses plant and terrestrial microbial communities in the Alaskan tundra. Chapter Three examines spontaneous stand regeneration and herb layer restoration in post-fire woods 16 years after a forest fire. Chapter Four reviews regularities and features of differentiation and anthropogenic transformation of steppe vegetation. In Chapter Five, the capacity of combination of biomass and native microorganism for fique bagasse from farmers from a region of Colombia named Oriente Antioqueño, was studied with the objective to degrade the tetracolorisoftalonitril active ingredient.

Chapter 1 - Saprotrophic microfungi were isolated from soil samples taken at different arctic ecosystems. Distinction of soil microfungi complexes taken at Arctic is shown. In the article, it is demonstrated that severe conditions of arctic area lead to formation of common characteristics for all complexes of soil microfungi, which are different from complexes of middle latitudes.

Complexes of microfungi in the Arctic regions are characterized by latitudinal differentiation from the polar desert zone to tundra zone (sub-zone arctic, typical and southern tundra) and the tundra-forest zone.

Adaptation to arctic natural conditions take place both at the level of a system, that is at the level of microfungi complexes, and at the level of species and isolates.

The general decreases of fungal numbers have been often noted in soil along the steep gradient directed from south to north.

In the polar desert, the species number decreases compared with more southern zones due to the poorness of soil's systematic content. Microfungi complexes at the polar desert show itself at the genus and species levels rather than at the level of more large-scale systematic groups.

In soil, the dynamics and mosaics of fungi distribution depend on diversity of microclimatic conditions at their sites of existence. Anthropogenic influence causes significant changes in the complexes of soil microfungi.

In the article, it is shown that location in Arctic climatic zones and plant-geographical zones lead to appearance (formation) of different characteristics of soil microfungi complexes i.e., general number, biomass, species and kinetic structure of communities, morphological, biochemical and functional features of microfungi, indexes of population densities and abundance of species at microfungi complexes.

Chapter 2 - The Alaskan tundra biome is a broad landscape largely distributed along the northern and western boundaries, occupying approximately 300,000 $mi^2$, or about one-third of the total land area (570,374 $mi^2$). It is part of the North American tundra biome located at latitudes 60° to 85° North and longitudes 55° to 160° West. The vegetation is largely dominated by moss, small shrubs, and some herbaceous plants. Climate change and global warming have made major impacts in Alaska. During the past 50 years, temperatures increased by an average of ~ 2.0°C, a rate that is twice the U. S. national average for the same period of time. Increasing temperatures and modest changes in precipitation have caused substantial thawing of the moss-rich permafrost, releasing massive amounts of soluble organic compounds that have accumulated over millennia in the form of frozen remains of peat moss that are deposited in strata each growing season. These soluble organic nutrients are utilized by soil microbes, thus increasing their metabolism, altering the composition of the microbial communities, and through increased respiration emitting substantially more $CO_2$ to the already, heavy-laden concentrations of global atmospheric $CO_2$. Concurrently, the ecology of Alaskan vegetation is undergoing marked changes, including transitions from largely moss and prostrate shrubs, to increasing invasion by small trees and other erect woody shrubs. Modern research techniques such as use of satellite data, experimental ecological field studies, and laboratory-based modeling and biogeochemical analyses of warming effects on plants and soil microbiota have improved the authors understanding of how climate change is affecting land plants and the soil microbial communities in the Alaskan tundra. Some current research evidence is summarized, including estimates of future effects on the environment and the consequences for human activity.

Chapter 3 - A survey of the development of the stand structure and the recovery of the herb layer in 16-year-old forests that appeared spontaneously after a forest fire was undertaken in southern Poland where the forests were burnt in 1992. Stand structure and its natural regeneration were surveyed in thinned and unthinned forests. The heights and diameters of trees were measured within each of the ten 40 x 20 m plots. In addition, ten transects (10 m long by 2 m wide, consisting of 5 quadrats, 4$m^2$ each) were set up in each of them. In each 2 x 2 m plot, the number of seedlings of tree and shrub species up to 0.5 m high, 0.5-2.0 m and over 2 m high that had a diameter < 5 cm were counted. The tree canopy cover was measured in each 2x2 m quadrat. Then, the herb layer composition in thinned and unthinned forests was studied. Thus, in randomly located 10x10 m plots (61 in the managed and 61 in unthinned forests), the percentage cover of herbaceous vascular plant species was estimated. Then, a numerical classification of the plots was conducted. The effect of thinning on stand composition and on the number of trees was very significant; the stand in thinned forest was exclusively composed of silver birch (density – 1,688-3,112 individuals per hectare), whereas in the unthinned forest the birch density reached 1,225-1,900 individuals per hectare and six other tree species were also present in this stand. Both forests differed in their canopy cover and the diameter of the birch trees (significantly higher in the thinned forests and lower in the unthinned forests, respectively). The response of a stand to thinning was also detectable in its regeneration process, which was more dynamic in the unthinned forest. A reaction to thinning was also noticed when the horizontal structure of the herb layer was compared. The most important difference was related to the increased level of the illumination of the forest floor within the thinned stand, which resulted in the increased abundance of different species of grasses and sedges, as compared to the herb layer in an unthinned stand. The main conclusion is that although there is a potential to provide timber of a better quality in the case of thinned forests, the recolonization of the herb layer by true forest species is slower than in successional forests that lack any management measures of their stands.

Chapter 4 - The questions of phytobiota evolution and formation of steppes in the vast region of the south of East European plain are reviewed. The main stages, aspects and ways of flora formation under increased anthropogenic impacts on steppe ecosystems are identified. Isolation and differentiation of different flora types at the present stage of development are shown. Their structure is determined by individuality of natural and anthropogenic interaction. The classification scheme of vegetation on the ecological-phytocenotic basis is developed and the vegetation types, formation classes, formations and associations are pointed out, using the results of long-term steppe phytocenoses research. The phytocenotic diversity of steppe vegetation in the systems of dominant and floristic classifications is characterized. The floristic richness, phytocoenotic diversity, uniqueness, stenotopic features, endemism, relictness, area-marginality of syntaxons are defined in the article. Rare for the region plant communities were identified. Regularities and peculiarities of the geographic and edaphic distribution of different level syntaxons of the south of East European plain steppe vegetation were established. It is shown that flora gene pool that has been preserved over centuries in the barrows and derelict lands can be used for eco-efficiency assessment of programs for the zonal steppe vegetation reconstruction.

Chapter 5 - In recent years the Colombian productive system has been characterized by a low level of technology and a high use of inputs. This dynamic is associated with the increased use of pesticides in rural zones, which has harmful effects on the soil and water and has repercussions for human and animal health. This affects the wetland ecosystems located in high Andean mountains; as such, reducing the concentration of these substances is an environmental priority.

2,4,5,6-Tetrachloroisophthalonitrile is one of the chemical substances most commonly used as the active fungicidal ingredient in fungicides applied to fruit, vegetable and ornamental crops.

On the other hand, the solid waste derived from the fique (Furcraea andina) agroindustry is an environmental problem. The extraction of fique fiber from the plants generates solid waste that is thrown into water sources.

This research studied the fungicide biodegradation potential of activity by microorganisms isolated from fique solid waste.

The isolated strains were incubated in order to determine which ones were resistant to environments rich in Tetrachloroisophthalonitrile. The effect of these microorganisms on the biodegradation of this compound (19 g $L^{-1}$) was evaluated over a 3-day period, with a maximum of 99.85% degradation occurring.

The in vitro study showed that isolated and purified organisms from fique solid waste can grow in and tolerate concentrations of fungicide, efficiently degrading it. This indicates a high potential for biodegradation of other chlorate substances.

In: Terrestrial Biomes
Editor: Marlon Nguyen
ISBN: 978-1-63484-625-7


*Chapter 1*

# GEOGRAPHIC DISTRIBUTION AND BIODIVERSITY OF MICROFUNGI IN SOILS OF THE ARCTIC REGIONS (AT TAIMYR PENINSULA AND NEARBY ISLANDS AS A SIMULATION MODEL OF ARCTIC)

***I. Yu. Kirtsideli and D. Yu. Vlasov****
V. L. Komarov Botanical Institute of the Russian
Academy of Sciences, St. Petersburg, Russia

## ABSTRACT

Saprotrophic microfungi were isolated from soil samples taken at different arctic ecosystems. Distinction of soil microfungi complexes taken at Arctic is shown. In the article, it is demonstrated that severe conditions of arctic area lead to formation of common characteristics for all complexes of soil microfungi, which are different from complexes of middle latitudes.

Complexes of microfungi in the Arctic regions are characterized by latitudinal differentiation from the polar desert zone to tundra zone (sub-zone arctic, typical and southern tundra) and the tundra-forest zone.

Adaptation to arctic natural conditions take place both at the level of a system, that is at the level of microfungi complexes, and at the level of species and isolates.

The general decreases of fungal numbers have been often noted in soil along the steep gradient directed from south to north.

In the polar desert, the species number decreases compared with more southern zones due to the poorness of soil's systematic content. Microfungi complexes at the polar desert show itself at the genus and species levels rather than at the level of more large-scale systematic groups.

In soil, the dynamics and mosaics of fungi distribution depend on diversity of microclimatic conditions at their sites of existence. Anthropogenic influence causes significant changes in the complexes of soil microfungi.

---

* Author for correspondence: Address: V. L. Komarov Botanical Institute of the Russian Academy of Sciences, 2 Prof. Popov St., St. Petersburg, 197376 Russia. E-mail: microfungi @mail.ru.

In the article, it is shown that location in Arctic climatic zones and plant-geographical zones lead to appearance (formation) of different characteristics of soil microfungi complexes i.e., general number, biomass, species and kinetic structure of communities, morphological, biochemical and functional features of microfungi, indexes of population densities and abundance of species at microfungi complexes.

**Keywords:** microfungi, soil, diversity, comparative ecology, microorganisms, moulds, polar desert, tundra, adaptation

## INTRODUCTION

Studies of various groups of organisms in extreme conditions are one of the most important problems of contemporary ecology. Developing of the organisms in extreme conditions always attracted scientists. Dependence of organisms and their communities on climatic factors is especially expressed in the arctic environment. One of the directions of studying organisms in extreme conditions is investigation of microfungi in soils of the Arctic and Antarctic.

Global warming (climatic change) with its consequences for weather and local climate, rising sea level, retreat of glaciers, etc. and possible nature catastrophes, actually is much disputed in newspapers, journals, and book publications (Kreisel 2006, 79; Bintanja and Linden 2013, 1557).

Polar region are cold, arid and windy. The mean air temperature of the warmest month in these regions is typically less than 10ºC, usually coinciding with the limits of tree growth (Legagneux et al. 2014, 379) and their terrestrial habitats are often snow and ice covered for several months of the year, with consequently short growing seasons. They are exposed to continuous sunlight or darkness for periods in summer and winter within the Arctic and Antarctic circles, parallels to latitude at 66º 33' north and south, respectively. Freeze-thaw events are usual. These harsh conditions impose strong selection pressures on plants inhabiting in Arctic and Antarctic ecosystems (Peat et al. 2007, 132).

The Arctic has a scarcity of biodiversity owing to extreme environmental conditions. Plant life of the region is scarce and restricted only to grasses, mosses, and lichens.

Mycological studies in the Arctic have focused on documenting mycorrhizal and herbaceous endophytes, lichenicolous fungi, and fungi at habitats such as soils, permafrost, deteriorated wood, ice, and marine waters at different location (Gilichinsky et al. 2005, 117-124; Pang et al. 2013, 5859-5862; Zhurbenko and Brackel 2013, 323).

Biogeography is a subject concerned with the limits and geometric structure of individual species populations and with the differences in biotas at various points on the earth's surface. The distribution of microfungi largely depends on the distribution of plant communities and nutrients (Fierer et al. 2009, 1238; Xu et al. 2013, 737; Ali et al. 2013, 39-52).

Traditionally, bio-geographers have accumulated information about the distribution of species, higher taxes, and the taxonomic composition of biota. Many kinds of microorganisms, including microfungi, are widespread and are found almost everywhere. This fact is based on the known principle stated by Beijerinck «Everything is everywhere» (Finlay 2002, 1061-63). Mycologists have only begun to apply the biogeographical theory in explaining fungal distribution patterns. Nowadays, the situation is changing and the data on geographic dependence is increased (Peay et al. 2010, 878-880; Jonsdottir 2011, 281; Nemergut et al. 2013, 342-353; Tedersoo et al. 2014, 1078-1015).

Biological diversity of soil microfungi in Arctic cenoses has not been studied sufficiently yet. This knowledge is important for understanding the functioning of ground-based ecosystems, especially in extreme conditions. High Arctic ecosystem may contain as much as 25-33% of the world's soil carbon (Mack et al. 2004, 440-441; Sistla et al. 2013, 615).

However, our knowledge of the community structure of saprotrophic microfungi, which is the important decomposers of an organic substance in extremely cold regions, is relatively scarce.

The cold climate leads that the tundra lives mainly underground (Bunnell and Scoullar 1981, 685). As a result, the interest to this soil inhabiting organisms increases. At the same time, conditions in the Arctic can vary over a wide range. Peculiarities of climatic conditions of the Polar region are their sharp latitudinal gradients. It is well noticeable in comparing of the average temperatures of the warmest month at different zones. On Taimyr tundra zone, the difference of temperatures of July throughout 700 km is 10°C (from 2° to 12°C), and on twice bigger territory of Central Siberia–only 6°C (from 12° to 18°C) (Callagan et al. 2004, 404-417; Chernov 2008, 342-345). As a result, it leads to a very big distinction both in communities of the highest plants and in other groups of organisms, including the complexes of soil microfungi.

Some research of soil microorganisms in the arctic regions began many years ago (Nystrom 1868, 551-571; Levin 1889, 558-567). These studies were not regular and had a reconnaissance nature. They had neither theoretical nor methodological bases for the assessment of microbiological population in this Arctic zone. Most researchers observed localization of microorganisms in the top soil horizons and with a sharp increase in the depth. Increase in the number of microorganisms was observed in the rizosphera of the arctic plants as compared with their number outside the roots zone (Kazanskiy 1932, 79-103).

Moreover, for a long time, the microbiological investigation carried out in the Arctic dealt with bacterial flora. In some cases, researchers observed the presence or absence of microfungi CFU in the soil. Data on specific (separated) location have been published in our articles (Kirtsideli 2009, 235-254; Zelenskaya et al. 2013, 135-141; Kirtsideli et al. 2014, 365-364) and some others. Comparison of data on the transect from polar deserts to the southern tundra of Taimyr peninsula is also shown in this article.

Soil samples were collected in this study from different ecosystem zones: 1) the polar desert - archipelago Northern Land (Severnaya Zemlya, mainly Bolshevik island) and island Vize (Wiese Island, Zemlya Vize) in Kara sea; 2) arctic tundra - Taimyr peninsulas, island Izvesiy TSIK (Kara sea); 3) typical tundra - Taimyr peninsula; 4) south tundra - Taimyr peninsula (Figure 1). Zonal division to zones of polar desert and tundra sub-zones is given by articles (Matveeva 1998, 56-78; Matveeva 2000, 219–224; Callaghan et al. 2013, 227-244). Similar border of arctic zones are provided in article (Walker et al. 2012, 1-15).

The microfungi were isolated using the soil dilutions plate techniques on agar medium. Identification was based on colony characteristics and microscopic features with the help of standard literature (Domsh at al. 2007, 23-780). Some species identification was carried out by molecular techniques.

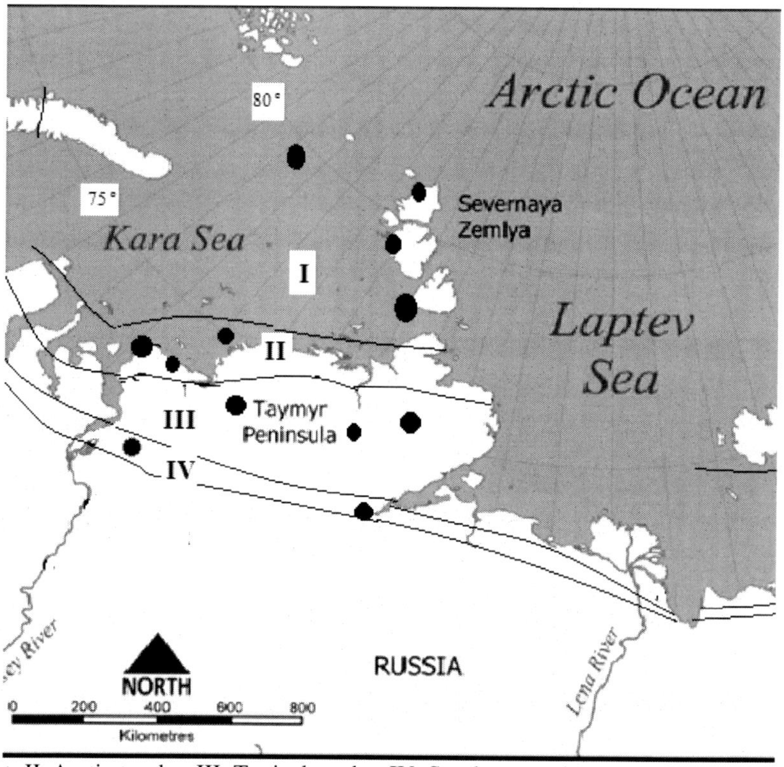


I–Polar desert; II–Arctic tundra; III–Typical tundra; IV–Southern tundra.

Figure 1. Map of the Arctic showing the area sampling.

## RESULT

The integral indexes of soil microfungi of the regions studied were always extremely low. Generally in our investigation, CFU numbers were less then 1000 CFU in 1 g of soil at polar desert (Figure 2), but in the tundra zone of arctic region this index may reach hundreds of thousands CFU. It is worth noting that although the CFU number oscillations are essential for underground cenoses, the number and biomass indexes may serve as indicators of polar desert and sub-zones of tundra.

The integral indexes of soil microfungi, or number CFU, isn't always considered unambiguously. Presence of microfungi spores in the soil isn't the evidence of their active participation in biological processes. However, this indicator can be considered as the test of potential biological activity of the soil processes. Within one zone and sub-zone, the number of soil microfungi depends on soil type, vegetable community, ecological conditions, and micro relief. As a rule, number of CFU is higher in intra-zonal communities in comparison with plakor habitats. The number CFU decreases under lichen curtains and increases in soils with a moss cover.

Under every zones and sub-zones changes the number indexes may vary in a wide range depending of cenoses. In addition, we have to note that microfungi number may strongly vary within one plant community. Thus in polar desert within the same plant community under the cushion of different plants, the number varied in a wide range (from 50 to 380 CFU in 1 g soil) where the most number were under the cushion of mosses (*Gymnomytrium corallioides, Racomitrium laniginosum*), and remained relatively high in soils under flower plants (*Aulacomnium turgidum, Deschampsia borealis*). The least numbers were noted in soils under lichens (*Cetraria encullata*) and in open soils. We may suppose that the open soil was subjected to more contrast temperature influence and the processes of freezing and thawing are take place more often there. In addition, the denser plant cover may serve as a buffer during temperature oscillation near zero level. Besides, under the cushion of plants, some stock of organic material are formed at the expense of death of vascular plants, and in the polar desert under the low content of organic material. This fact may be a determining one.

Anthropogenic and zoogenic influences to the soil may change microfungi number in cenoses. Northern soils are usually poor in nitrogen (Robinson et al. 2004, 1066). In zoogenic communities enriched by this element, the number of microfungi is increased in the soil. At polar desert soils of Bolshevic island (Northern Land) influenced by zoogenic impact, the number of soil microfungi increases to 180-215 CFU. In upland (plakor) cenosis CFU number doesn't exceed 150 CFU in 1 g soil. At arctic tundra soils (island Izvesiy TSIK) influenced by zoogenic, impact the number of soil microfungi reached to 3500 CFU from 1500 CFU at upland cenosis (plakor).

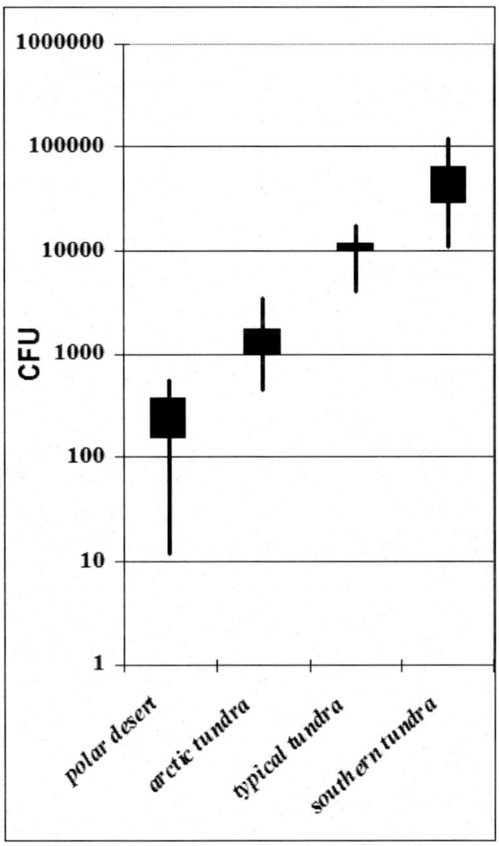


Figure 2. The numbers of microfungi (CFU) in soils at polar desert and tundra sub-zones (rectangle– CFU number at plakor soils).

At polar desert soils of Bolshevic island the most CFU number (220-250 CFU per 1 g of air-dry soil) was noted in community subjected to anthropogenic influence. The anthropogenically contaminated soils of the island Vize (polar desert), the number of microfungi increased to 560 CFU (from 370 CFU at upland (plakor) cenosis).

At arctic tundra on island Izvesiy TSIK, anthropogenically contaminated soils contained 1100 CFU; soils of upland cenosis (plakor) contained 1500 CFU. At arctic tundra on Taimyr Peninsula (near Dicson Settelment), anthropogenically contaminated soils contained 1900 CFU; soils of upland cenosis (plakor) contained 1200 CFU. At typical tundra on Taimyr Peninsula, anthropogenically contaminated soils contained 17250 CFU; soils of upland cenosis (plakor) contained 12000 CFU.

The influence of temperature to Arctic microfungi development in laboratory conditions may be non monosemantic and depend from ecological specific. Thus, when studying microfungi number CFU index at polar desert after incubation at 10°C, was higher compared with that at 20°C. In plant communities developing on site of a melted snow patch, considerable increase of number was noted at 5°C temperature. The difference in microfungi number incubated under different temperatures may be explained by the fact that at 20°C, microfungi with high growth rate suppress the slowly growing species and the latter could not be visually fixed and isolated as culture. Besides, the incubation temperature above 20°C could suppress spores germination of some psychrophilic species.

## The Activity of Soil Microfungi

The method of soil dilutions on agar medium gives information on total amount of microbiota which is in active and passive conditions. The use of the "growing on glass" method permits us to characterize microfungi complexes in specific conditions. It was noted that over a comparatively shot period of exposition (less than 1 month), mycelium appeared on glass surfaces, which evidences for comparatively high biological activity of soil microfungi. Microfungi form different mycelium considering sizes, branching, and colours. Microfungi on glasses did not form any conidial in polar desert and Arctic tundra (this must be connected with a comparatively short period of exposition). This method showed distinctions in the square of a projective covering at exhibiting in polar deserts and tundra sub-zones (Figure 3). This method can serve as the indicator of change of climatic and vegetable zones at advance from the North to the South from a zone of polar deserts to the southern tundra.

## Taxonomical Structure of Microfungi from Polar Desert

At polar deserts soils, microfungi complexes amounted to 62 species. At arctic tundra, soils microfungi complexes amounted to 95 species. At typical tundra soils, microfungi complexes amounted to 104 species. At southern tundra soils, microfungi complexes amounted to 121 species. At Arctic territory soils (At Taimyr Peninsula and nearby islands), microfungi complexes amounted to 193 species from 69 genera. The number of species in genera varied from 1 to 48. Zygomycota totaled 23 species and Basidiomycota 6 species. All other species belong to Ascomycota.

Some isolates of mycelium sterile were not identified and considered here as mycelium sterile group of light and dark coloured types, part of which is possibly Basidiomycota.

It is worth noting that information on the species content is not complete on account of some restriction in the sites studied and cenoses types since the species with low occurrence or/and with low population density are isolated only under considerable number of recurrences. In cenoses subjected to anthropogenic and zoogenic influence the number of scarce species increases and decrease of typical species takes place. That considerable increased the general list of species.

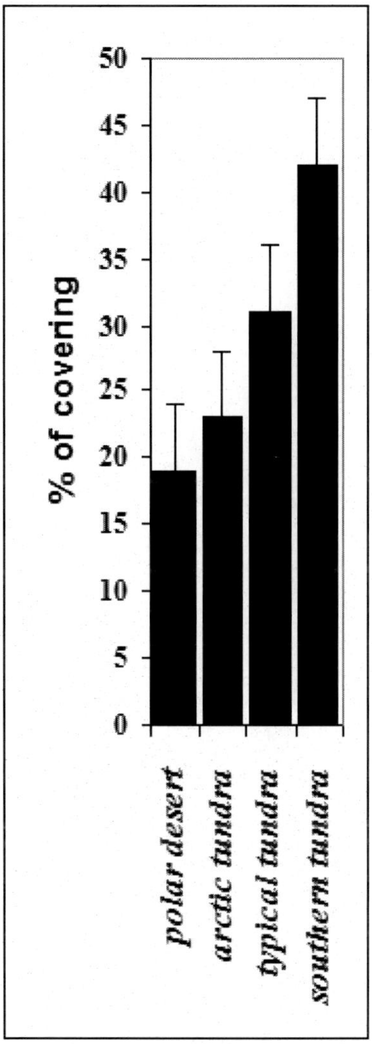


Figure 3. Area of a projective covering microfungi on glasses during the incubation in soils at polar deserts and tundra sub-zones.

# Structure of Soil Microfungi Complexes

Analysis of frequency index of genera, as well as the index of population density, demonstrated change of dominant genus in microfungi complexes from polar desert to south tundra (Figure 4). Value of the population density of basic microfungi groups has very specific character in different zones of the Arctic (in soils of polar desert and subzones tundra).

As a rule, the percentage of microfungi of the genus *Geomyces* was very great in the polar desert soils, where the proportion of species *G. pannorom* sometimes reached 70% (56% mean). The portion of Zygomycota was relatively low. In soils of arctic tundra subzones, Geomyces pannorum portion decreased to 26%, and its portion decreased at typical and south tundra to 11 and 12%.

Genus *Penicillium* was represented by 1-4 species at soil of polar desert and its portion was near 5% and increased to 27% at arctic tundra and to 39% and 45% at typical and south tundra. The number of species of this genus also increased from 6 at polar desert to 42 in the southern tundra. A similar trend was noted for Zygomycota species. Their proportion increased from polar deserts to the southern tundra.

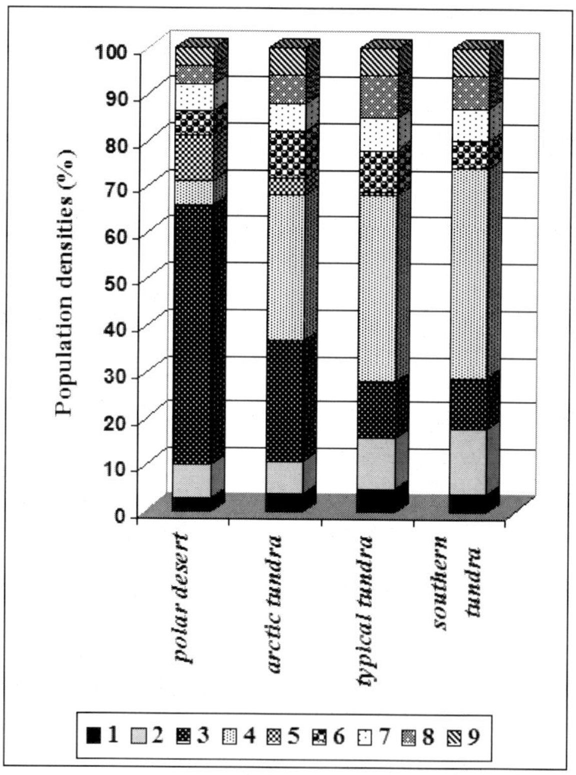


1–Basidiomycota; 2–Zygomycota; Ascomycota with orders: 3–Helotiales (including genus Geomyces, Cadophora), 4–Eurotiales (including genus Penicillium); 5–Thelebolales; 6–Hypocreales; 7–Pleosporales; 8–Capnodiales; 9–all other Ascomycota.

Figure 4. Population densities of the main group of microfungi in soils at polar desert and tundra subzones.

Structure of soil microfungi complexes at polar desert is very specific. The oligo-domination principle can be clearly observed in this case. Poorness of taxonomical structure (systematic content) of microfungi in Arctic soils was noted both at the species and the genus levels. The more large-scale systematic units represented completely. For instance, such prevalent usual for southern soils genera as *Absidia, Gliocladium*, etc. were not observed at polar desert and arctic tundra. At polar desert isolated as random the species of genera *Fusarium, Trichoderma* were which are considered as typical ones for middle latitude soils.

The numerical structure of soil microfungal complexes of the polar desert is different from analogous indices for tundra zone soils. Almost all species noted by us for polar desert cenoses are also known for plain soils of the more south zone. Microfungi from genus *Antarctomyces* and *Thelebolus* are expception.

Physiological and biochemical features polar desert populations (psychrophiles) allow them to successfully function at extreme cold conditions. Moist microfungi isolated from the soils of polar desert have temperature growth optimum below 20°C (10°-15°C) which is higher than conditions of their existence in nature, but at the same time, most of the isolates are able to grow and develop under lower temperatures. High quota of some species account most likely by their psychrophiles characteristics.

Spores of the majority of isolated species are able almost not to lose their vital capacity under negative temperatures during a long period time. It is interesting to note that the microfungi not creating spores (mycelium sterile) isolated from arctic soil did not lose their vitality in soil under negative temperatures either. Apparently, most species of microfungi isolated from polar desert soil are able to develop only during short Arctic summer and only small portion of species are able to develop more long period. Probably that may determine long duration of the processes of biological destruction of organically material in polar desert.

## CONCLUSION

The mycobiota of Arctic soil cenoses is characterized by relatively little list species. The poorness of the systematic content of soil microfungi complexes of the polar desert and tundra sub-zones shows itself at the level of species and genera. The more large-scale systematic units represented completely. The total number of soil microfungi at Arctic is always low. The limit of the total number and biomass of microfungi in one zone are very similar and may be use as indicators of polar desert and subzones of tundra.

The plant cover is one of the main factors to determine microfungi number in soils of one zone and tundra sub-zones, in some cases it seems to be a key position(s) (limitation factor) of microfungi development. Zoogenic and anthropogenic influence by organic pollution on soils may be stimulating factors for soil microfungi.

In laboratory condition (*in-vitro*) temperature is one of selecting factor in determination of microfungi number. Characteristic feature of mycobiota of polar desert seems to be the oligo-domination of soil microfungi complexes, wide ecological limit and psychrophiles of species.

Climate change from polar deserts to sub-zones of the southern tundra leads to change the complex structure of microfungi as well as variety in proportion of different species and genus. Comparison of number, biodiversity, and structure of microfungi complexes is necessary to conduct only on plakor cenoses. Intra-zonal cenoses, zoogenic, and anthropogenic influence (organic pollution) change these parameters.

## ACKNOWLEDGMENT

This works were supported in part from the Russian Found for Basic Research.

## REFERENCES

Ali, S. H., Alias, S. A., Siang, H. Y., Smyklla, J., Pang, K. L., Guo, S. Y. and Convey, P. 2013. "Studies on diversity of soil microfungi in the Hornsund area, Spitsbergen." *Pol. Polar Res.* 34(1): 39-54.

Bunnell, F. L. and Scoullar, K. A. 1981. "Between-site comparisons of carbon flux in tundra by using simulation models." In: *Tundra ecosystems: A comparative analysis*, by L. C. Bliss, O. W. Heal and J. J. Moore, 685-715. Cambridge, UK: Cambridge Univ. Press.

Domsch, K. H., Gams, W. and Anderson, T. H. 2007. *Compendium of Soil Fungi.* Second edition. IHW-Verlag Eching.

Bintanja, Richard and van der Linden, E. C. 2013. "The changing seasonal climate in the Arctic." *Nature Scientific Reports* 3:1556-68.

Callagan, T. V., Bjorn, L. O., Chernov, Yu. I., Chapin, T., Christensen, T. R., Huntlet, B., Johansson, M., Jolly, D., Jonasson, S., Matveyeva, N., Panikov, N., Oechel, W., Shaver, G., Elster, J., Henttonen, H., Laine, K., Taulavuori, K., Taulavuori, E. and Zockler, C. 2004. "Biodiversity, distribution and adaptations of Arctic species in the context of environmental change." *Ambio.* 33(7): 404-17.

Callaghan, T. V., Matveyeva, N., Chernov, Yu., Schmidt, N. M., Brooker, R. and Johansson, M. 2013. "Arctic Terrestrial Ecosystems" by Levin S. A. (ed.) *Encyclopedia of Biodiversity*, second edition. Vol. 1. 227-244. Waltham, MA: Academic Press.

Chernov, Yu. I. 2008. *Ecology and Biogeography.* Moscow: KMK (in Russian).

Fierer, N., Strickland, M. S., Liptzin, D., Bradford, M. A. and Cleveland, C. C. 2009. "Global patterns in belowground communities." *Ecol. Lett.* 12: 1238-49.

Finlay, B. J. 2002. "Global dispersal of free-living microbial eukaryote species." *Science*, 296: 1061-63.

Gilichinsky, D., Rivkina, E., Bakermans, B., Shcherbakova, V., Petrovskaya, L., Ozerskaya, S., Ivanushkina, N., Kochkina, G., Laurinavichuis, K., Pecheritsina, S., Fattakhova, R. and Tiedje J. M. 2005. "Biodiversity of cryopegs in permafrost." *FEMS Microbiology Ecology* 53: 117-28.

Jonsdottir, I. S. 2011. "Diversity of plant life histories in the Arctic." *Preslia* 83: 281-300.

Kasansky, À. F. 1932. "About microfungi from New Land." *Òr. polar commission ÀN USSR.* 7: 79-108 (in Russian).

Kirtsideli, I. Yu. 2009. "Soil microfungi from polar deserts." *Species and Communities in Extreme Environments.* 235-254. Pensoft Publishers and KMK Scientifi c Press. Sofia-Moscow.

Kirtsideli, I. Yu., Vlasov, D. Yu., Barantsevich, E. P. and Sokolov, V. T. 2010. "Microfungi from soil of polar desert at Izvestia island (in Kara sea)." *Mikologiya i Fitopatologiya* 44(2): 365-67 (in Russian).

Kreisel, H. 2006. "Global warming and mycoflora in the Baltic Region." *Acta Mycologica* 41:79-94.

Legagneux, P., Gauthier, G., Lecomte, N., Schmidt, N. M., Reid, D., Cadieux, M-C., Berteaux, D., Bêty, J., Krebs, C. J., Ims, R. A., Yoccoz, N. G., Morrison, R. I. G., Leroux, S. J., Loreau, M. and Gravel, D. 2014. "Arctic ecosystem structure and functioning shaped by climate and herbivore body size." *Nature Climate Change* 4:379-83.

Levin, C. 1889. "Les microbes dans les regions Arctiques." *Ann. Inst. Paster.* 13:558-67.

Mack, M. C., Schuur, E. A. G., Bret-Harte, M. S., Shaver, G. R. and Chapin, F. S. 2004. "Ecosystem carbon storage in arctic tundra reduced by long-term nutrient fertilization." *Nature.* 431: 440-43.

Matveeva, N. V. 1998. *Zonality in Arctic vegetation cover.* S.-Petersburg: Nauka (in Russian).

Matveyeva, N. V. 2000. "*Taymyr as mini-Arctic model.*" Heritage of the Russian Arctic research, conservation and international co-operation. Moscow: Ecopros Publishers (in Russian).

Nemergut, D. R., Schmidt, S. K., Fukami, T., O'Neill, S. P., Bilinski, T. M., Stanish, L. F., Knelman, J. E., Darcy, J. L., Lynch, R. C., Wickey, P. and Ferrenberg, S. 2013. "Patterns and processes of microbial community assembly." *Microbiol. Mol. Biol. Rev.* 77:342-56.

Nystrom, C. 1868. "Om jashings-och forruttnel seprossena pa Spetsbergen." *Uppsala lakarrforenings Forrhandingar.* 4:551-71.

Pang, K. L., Chow, R. K. K., Chan, C. W. and Vrijmoed, L. L. P. 2011. "Diversity and physiology of marine lignicolous fungi in Arctic waters: a preliminary account." *Polar Research* 30:5859-65.

Peat, H. J., Clarke, A. C. and Convey, P. 2007. "Diversity and biogeography of the Antarctic flora." *Journal of Biogeography.* 34:132-46.

Peay, K. G., Bidartondo, M. I. and Arnold, A. E. 2010. "Not every fungus is everywhere: Scaling to the biogeography of fungal-plant interactions across roots, shoots and ecosystems." *New Phytol.* 185:878-82.

Robinson, C. H., Saunders, P. W., Madan, N. J., Pryce-Miller, E. J. and Pentecost, A. 2004. "Does nitrogen deposition affect soil microfungal diversity and soil N and P dynamics in a high Arctic ecosystem?" *Global Change Biology.* 10(7):1065-79.

Sistla, S. A., Moore, J. C., Simpson, R. T., Gough, L., Shaver, G. R. and Schimel, J. P. 2013. "Long-term warming restructures Arctic tundra without changing net soil carbon storage." *Nature.* 497:615-18.

Tedersoo, L., Bahram, M., Põlme, S., Kõljalg, U., Yorou, N. S., Wijesundera, R., Ruiz, L. V., Vasco-Palacios, A. M., Thu, P. Q., Suija, A., Smith, M. E., Sharp, C., Saluveer, E., Saitta, A., Rosas, M., Riit, T., Ratkowsky, D., Pritsch, K., Põldmaa, K., Piepenbring, M., Phosri, C., Peterson, M., Parts, K., Pärtel, K., Otsing, E., Nouhra, E., Njouonkou, A. L., Nilsson, R. H., Morgado, L. N., Mayor, J., May, T. W., Majuakim, M., Lodge, D. J., Lee, S. S., Larsson, S. H., Kohout, P., Hosaka, K., Hiiesalu, I., Henkel, T. W., Harend, H., Guo, L., Greslebin, A., Grelet, G., Geml, J., Gates, J., Dunstan, W., Dunk, C., Drenkhan, R., Dearnaley, J., De Kesel, A., Dang, T., Chen, X., Buegger, F., Brearley, F. Q., Bonito, G.,

Anslan, S., Abell, S. and Abarenkov, S. 2014. "Global diversity and geography of soil fungi." *Science*. 346:1078-1125.

Walker, D. A., Epstein, H. E., Raynolds, M. K., Kuss, P., Kopecky, M. A., Frost, G. V., Dani¨els, F. J. A., Leibman, M. O., Moskalenko, N. G., Matyshak, G. V., Khitun, O. V., Khomutov, A. V., Forbes, B. C., Bhatt, U. S., Kade, A. N., Vonlanthen, C. M. and Tich´y, L. 2012. "Environment, vegetation and greenness (NDVI) along the North America and Eurasia Arctic transects." *Environ. Res. Lett.* 7:1-17.

Xu, X., Thornton, P. W. and Post, M. 2013. "A global analysis of soil microbial biomass carbon, nitrogen and phosphorus in terrestrial ecosystems." *Glob. Ecol. Biogeogr.* 22:737-49.

Zelenskaya, M. S., Kirtsideli, I. Yu., Vlasov, D. Yu., Krylenkov, V. A. and Sokolov V. T. 2013. "Micromycetes-biodestructors in the Arctic geobiocenosis." *Regional environmental issues*. 5:135-141 (in Russian).

Zhurbenko, M. P. and Brackel, W. 2013. "Checklist of lichenicolous fungi and lichenicolous lichens of Svalbard, including new species, new records and revisions." *Herzogia* 26:323-59.

## Biographical Sketches

### *Kirtsideli I. Ju.*

*Name*: Kirtsideli I. Ju.

*Affiliation*: V. L. Komarov Botanical Institute of the Russian Academy of Sciences

*Date of Birth*: 06/10/1962

*Education*: M.Sc.–1985, Saint-Petersburg State University, Russia/ Department of Biology and Soil Biology/biology (Botany),

Ph.D.–1993. Mycology at V. L. Komarov Botanical Institute of Russian Academy of Sciences

*Address*: Prof. Popov St. 2, 197376 St. Petersburg, Russia

tel/+7-921-5784128

E-mail: microfungi@mail.ru

*Research and Professional Experience*:

- microfungi in extreme conditions, structural and functional organization of microbial associations in natural and industrial soils;
- biodiversity of soils microorganisms;
- decomposition organic matter in soil;
- restoration of biological properties of soils after its chemical pollution;
- ecology, population biology, number, biomes, structure of complexes and productivity of microfungi.
- Morphology and morphogenesis of fungi and adaptive role of morphogenesis. Influence of extreme signals on fungal morphogenesis. Experimental study of fungi.

*Professional Appointments*:

Since 1998 Senior Researcher, Laboratory of fungal taxonomy and geography, V. L. Komarov Botanical Institute of Russian Academy of Sciences.

1993-1998 Junior research worker, Laboratory ecology of fungi. Komarov Botanical Institute of Russian Academy of Sciences.

1985-1990 Junior Researcher in the Laboratory ecology of fungi, Komarov Botanical Institute of Russian Academy of Sciences St. Petersburg, Russia.

Participant of 58 Russian Antarctic Expedition (2012-13) and Arctic Expedition 1984-1986, 1993-2000, 2003-2007, 2010-2011.

*Some publications Last Three Years*:

Distribution of terrigenousterrigene microfungi in Arctic seas. *Mycology and Phytopathology*. 2012. 46(5): 306-310 (in Russ.).

Kirtsideli, I. Yu., Vlasov, D. Yu., Barantsevich, E. P., Krylenkov, V. A., Sokolov, V. T. Microfungi from soil of polar island IzvestaTsik (Kara sea). *Mycology and Phytopathology*. 2014. 48 (6): 425-431 (in Russ.).

Sazanova, K. V., Kirtsideli, I. Yu. Influence of UV irradiation on microfungi? isolated from Antarctic habitat. *Mycology and Phytopathology*. 2014. 48 (5): 315-321 (in Russ.).

Ukhanova, O. P., Bogomolova, E. V., Kirtsideli, I. Yu. Dynamics of the content of propagules potentially allerge nic microscopic fungi in the air of St Petersburg. *Allergology and Immunology*. (Official Journal of the CIS Society of Allergology and Immunology) 2013. 14(3): 174-178 (in Russ.).

Vlasov, D. Yu., Zelenskaya, M. S., Kirtsideli, I. Yu., Abakumov, E. V., Krylenkov, V. A., Lukin, V. V. Fungi on the natural and anthropogenic substrates in West Antarctica. *Mycology and Phytopathology*. 2012. 46(1): 20-26 (in Russ.).

Zelenskaya, M. S., Kirtsideli, I. Yu., Vlasov, D. Yu., Krylenkov, V. A., Sokolov, V. T. Micromycetes – biodestructors of the Arctic ecosystems. *Problems of regional ecology*. 2013. 5: 135-141 (in Russ.).

## *Dmitry Vlasov*

*Name*: Vlasov Dmitry

*Affiliation*: St. Petersburg State University

Botanical Institute of Russian Academy of Science

*Date of Birth*: 11/09/1959

*Education*: Diploma, 1981, Saint-Petersburg State University, Russia/ Department of Biology and Soil Biology/biology

Ph.D., 1987, Plant Protection, All-Union Research Plant Protection Institute/Russia

Dc. Sc., 2008, Mycology, Botanical Institute of Russian Academy of Science/Russia

*Address*: Saint Petersburg State University, Biology and Soil Faculty, Botany Department.

199034, Russia, Saint-Petersburg, Universitetskaya nab., 7/9,

Tel: +7 9117109428

Fax: +7 812 4507310

E-mail: Dmitry.Vlasov@mail.ru

*Research and Professional Experience*:

Ecology, adaptation and evolution of microscopical fungi (micromycetes), microbial communities in the extreme environment, microbial deterioration and degradation of stone, geomicrobiology, biodeterioration of natural and anthropogenic substrates, mechanisms of biodeterioration, ecological aspects of biodeterioration, ecology of fungi, polar research.

*Teaching* (Saint-Petersburg State University).

*Lecturing*: Mycology, Algology, Ecology of fungi, Geomicrobiology, Soil mycology, Common phytopathology;

*Practice*: fungi, alga, lichens.

Supervisor of 5 PHD works (successfully protected). Supervisor of 18 graduate works.

*Professional Appointments*:

1. All-Russian Plant Protection Institute, Lab. Mycology, (1987-1991)
   - Scientist,
   - Leader of the scientific group on soil fungi investigation.
2. Biological Research Institute of Saint Petersburg State University.
   - Chief of the Lab. Mycology (1991-1999).
   - Vice-director on scientific work (1999-2007).
3. Saint Petersburg State University, Biology and Soil Faculty
   - Head of the Dep. of Scientific work of the Biology and Soil Faculty (2007-2009)
4. Saint Petersburg State University, Biology and Soil Faculty, Dep. of Botany:
   - Professor (the present time)
5. Botanical Institute of Russian Academy of Science:
   - Leading Researcher (the present time)
6. Participant of 49, 52, 54, 56 Russian Antarctic Expedition (2004, 2006, 2008, 2010)

*Honors*:

"Leonard Eyler Medal" (European Academy of Natural Sciences), 2014.

*Publications Last Three Years*:

Abakumov, E. V., Gagarina, E. I., Sapega, V. F., Vlasov, D. Yu. Micromorphological features of the fine earth and skeletal fractions of West Antarctica soils in the areas of Russian Antarctic stations. *Eurasian Soil Science*, 2013, Vol. 46, No. 12, pp. 1219-1229.

Barinova, K. V., Vlasov, D. Yu., Schiparev, S. M. The influence of zinc and cooper on acid production by fungus Penicillium citrinum *in vitro. Mycology and Phytopathology*. 2012. V. 46, N 6. P. 385-390.

Dashko, R. E., Vlasov, D. Yu., Shidlovskaya, A. V. Geotechnics and subsurface microbiota. *SPb.: Georekonstruktsiya*, 2014. 280p. (In Russian).

Frank-Kamenetskaya, O. V., Vlasov, D. Yu. *Monitoring of the stone monuments*. St. Petersburg: The Institute of Earth Sciences, 2014, 60p (In Russian).

Frank-Kamenetskaya, O. V., Vlasov, D. Yu. *Monitoring of rock monument state*. SPb.: SPbSU, 2014, 32 p. (In Russian).

Frank-Kamenetskaya, O., Rusakov, A., Barinova, K., Zelenskaya, M., Vlasov, D. The formation of oxalate patina on the surface of carbonate rocks under influence of microorganisms. *Applied Mineralogy* (ICAM-2011). Maarten A.T.M. Broekmans (ed.) Trondheim, Berlin Heidelberg: Springer-Verlag, 2012, p. 213-220.

Frank-Kamenetskaya, O. V., Vlasov, D. Yu., Shilova, O. A. Biogenic Crystals Genesis on a Carbonate Rock Monument Surface: The Main Factors and Mechanisms, the Development of Nanotechnological Ways of Inhibition. *Minerals as advanced Materials* II. Springer-Verlag, 2012. P. 401-413.

Khamova, T. V., Shilova, O. A., Vlasov, D. Yu., Ryabusheva, Yu. V., Mikhal'chuk, V. M., Ivanov, V. K., Frank-Kamenetskaya, O. V., Marugin, A. M., Dolmatov, V. Yu. Bioactive Coatings Based on Nanodiamond Modified Epoxy Siloxane Sols for Stone Materials. *Inorganic Materials*. 2012. Vol. 48, No. 7. P. 702-708.

Kirtsideli, I. Y., Vlasov, D. Yu., Barantsevich, E. P., Krylenkov, V. A., Sokolov, V. T. Distribution of terrigenous micromycetes in waters of the Arctic seas. *Mycology and Phytopathology*. 2012. V. 46, N 5. P. 306-310 (In Russian).

Kozlowski, A., Frank-Kamenetskaya, O. V., Chelibanov, V. P., Abakumov, E. V., Vlasov, D. Yu. Sulfur cycle in industrial megalopolises due to its influence on the state of monuments of marble and limestone. *Problems of regional ecology*. 2013. № 5. P. 172-178 (In Russian).

Panin, A. L., Bogumilchuk, E. A., Sharov, A. N., Vlasov, D. Yu., Zelenskaya, M. S., Tolstikov, A. V., Teshebaev, S. B., Tseneva, G. Y., Sboychakov, V. B., Bolehan, V. N. Microbial mat as objects for monitoring the Antarctic ecosystem. Vestnik St. Petersburg University. *Series 3: Biology*, 2013. No 2. P. 3-11 (In Russian).

Panova, E., Vlasov, D., Luodes, H. Evaluation of the durability of granite in architectoral monuments. *Printing house: Juvenes Print – Tampereen yliopistopaino Oy, Espoo,* 2014. 90p.

Popova, T. A., Vlasov, D. Yu., Zelenskaya, M. S., Panova, E. G. Biofouling of granite embankments in Saint Petersburg. *Vestnik of SPbSU, Biology*, 2014 (2): 30-40 (In Russian).

Sazanova, K., Vlasov, D., Osmolovskaya, N., Schiparev, S., Yakkonen, K., Kuchaeva, L. Organic Acids Induce Tolerance to Zinc- and Copper-Exposed Fungi Under Various Growth Conditions. *Curr. Microbiol.* (2015) 70:520-527 (DOI 10.1007/s00284-014-0751-0).

Sazanova, K. V., Schiparev, S. M., Vlasov, D. Yu. Formation of Organic Acids by Fungi isolated from the Surface of Stone Monuments. *Microbiology*, 2014, Vol. 83, N 5, P. 516-522.

Vlasov, D. Yu., Zelenskaya, M. S., Kirtsideli, I. Y., Abakumov, E. V., Krylenkov, V. A., Lukin, V. V. Fungi on natural and anthropogenic substrates in Western Antarctica. *Mycology and Phytopathology*, 2012. V. 46, N 1. P. 20-26 (In Russian).

Vlasov, D. Yu., Tobias, A. V., Cherepanova, N. P. Mycology development on the Botany Department: traditions and new directions. *Vestnik of SPbSU*, 2013. *Ser. 3 Biology*. N 3, P.135-149 (In Russian).

Vlasov, D. Yu., Abakumov, E. V., Tomashunas, V. M., Krylenkov, V. A., Zelenskaya, M. S. Mycobiota of soil and anthropogenic substrates of the Yamal Peninsula. *Gigiena i Sanitaria*. 2014. 2014, № 5 C. 49-51 (In Russian).

Zelenskaya, M. S., Kirtsideli, I. Yu., Vlasov, D. Yu., Krylenkov, V. A., Sokolov, V. T. Micromycetes-biodestructors of the Arctic ecosystems. *Problems of regional ecology*. 2013, № 5. P. 135-141 (In Russian).

In: Terrestrial Biomes
Editor: Marlon Nguyen

ISBN: 978-1-63484-625-7


*Chapter 2*

# THE ALASKAN TUNDRA: PLANT AND TERRESTRIAL MICROBIAL COMMUNITIES IN A CHANGING CLIMATE

***O. Roger Anderson***
Biology, Lamont-Doherty Earth Observatory of Columbia University, Palisades, NY, US

## ABSTRACT

The Alaskan tundra biome is a broad landscape largely distributed along the northern and western boundaries, occupying approximately 300,000 $mi^2$, or about one-third of the total land area (570,374 $mi^2$). It is part of the North American tundra biome located at latitudes 60° to 85° North and longitudes 55° to 160° West. The vegetation is largely dominated by moss, small shrubs, and some herbaceous plants. Climate change and global warming have made major impacts in Alaska. During the past 50 years, temperatures increased by an average of ~ 2.0°C, a rate that is twice the U. S. national average for the same period of time. Increasing temperatures and modest changes in precipitation have caused substantial thawing of the moss-rich permafrost, releasing massive amounts of soluble organic compounds that have accumulated over millennia in the form of frozen remains of peat moss that are deposited in strata each growing season. These soluble organic nutrients are utilized by soil microbes, thus increasing their metabolism, altering the composition of the microbial communities, and through increased respiration emitting substantially more $CO_2$ to the already, heavy-laden concentrations of global atmospheric $CO_2$. Concurrently, the ecology of Alaskan vegetation is undergoing marked changes, including transitions from largely moss and prostrate shrubs, to increasing invasion by small trees and other erect woody shrubs. Modern research techniques such as use of satellite data, experimental ecological field studies, and laboratory-based modeling and biogeochemical analyses of warming effects on plants and soil microbiota have improved our understanding of how climate change is affecting land plants and the soil microbial communities in the Alaskan tundra. Some current research evidence is summarized, including estimates of future effects on the environment and the consequences for human activity.

# INTRODUCTION

Geographically, Alaska is situated at the northern-most part of the North American continent, with a geographic center 63° 50' north latitude, 152° west longitude, approximately 60 miles northwest of Denali; and occupies 586,412 square miles of land area - a total area one-fifth the size of the lower 48 states. Based on landforms, Alaska can be divided into four general physiographic regions (Figure 1): Arctic Coastal Plain (Interior Plains), Rocky Mountain System of Alaska, Central Uplands and Lowlands (Intermontane Basins and Ranges), and Pacific Mountains and Valleys. Further designated regions by latitude (north to south) include the Arctic Coastal Plain, North Slope, Brooks Mountain Range, a central upland dissected by the Yukon River, the massive Alaska Mountain Range, the Pacific Coastal areas and eastern Inside Passage, and the Alaskan Peninsula, and Aleutian Islands of the southwest.

The Alaskan tundra is a broad landscape largely distributed along the northern and western boundaries in the Arctic Coastal Plain, occupying approximately 300,000 $mi^2$, or about one-third of the total land area (570,374 $mi^2$). It is part of the North American tundra biome located at latitudes 60° to 85° North and longitudes 55° to 160° West. It lies north of the Rocky Mountain System and slopes gradually toward the Arctic Ocean.

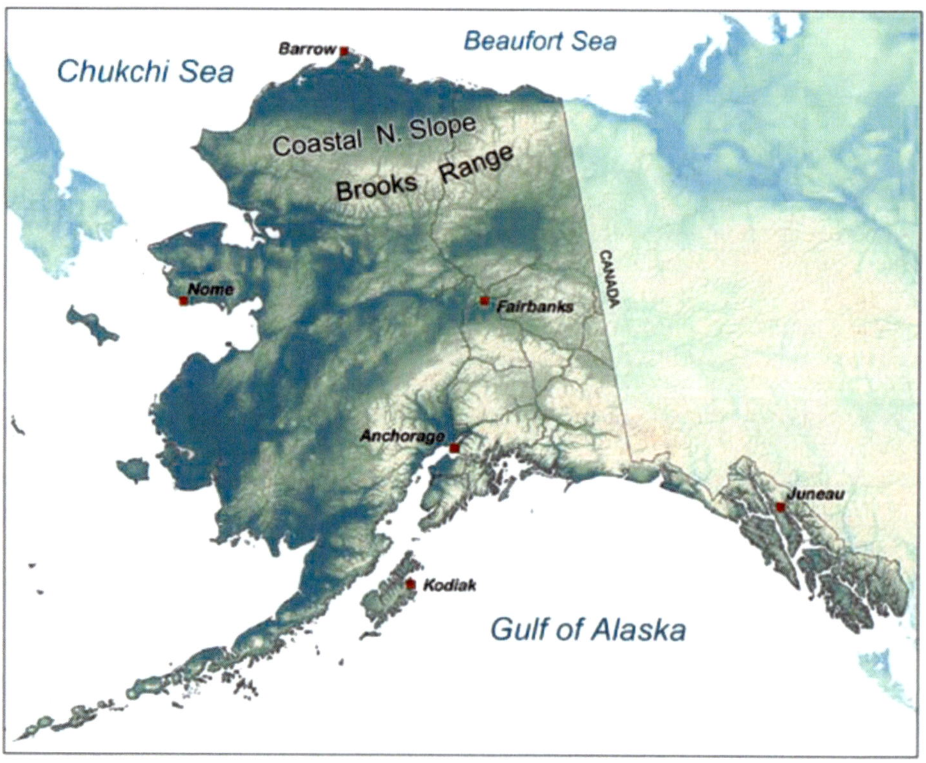


Figure 1. Map of Alaska showing the mountains of the Brooks Range and Coastal N. Slope plains, where extensive tundra is located. Adapted from a State Government of Alaska image (http://www.adfg.alaska.gov/index.cfm?adfg=viewinglocations.main).

The Brooks Range Mountains in northern Alaska is a northern extension of the Rocky Mountains. It reaches 600 mi westward from Canada to the Chukchi Sea. Its rugged peaks reach elevations of 9,000 ft in the east, falling to 3,000 ft in the west. A series of rolling plateaus and low mountains, the arctic foothills, borders the coastal plain to the north forming a "polar tundra desert" characterized by a moss-dominated vegetation, and moist tundra-cottongrass tussocks, interspersed with willow-dominated communities along river corridors. Wetlands occupy greater than 83% of the area. It has been a geographic site for substantial recent ecological and climate change research, including the Arctic Long-term Ecological Research (LTER) site located at Toolik Lake (e.g., http://ecosystems.mbl.edu/ARC/) on the North Slope. The Toolik Lake area is characterized by continuous permafrost, no trees, a complete snow cover for seven to nine months annually, winter ice cover on lakes, streams, and ocean, and cessation of river flow during the winter. Tussock tundra is the dominant vegetation type, but there are extensive areas of drier heath tundra on ridge tops and other well-drained sites as well as areas of river-bottom willow communities.

Alaska has been dubbed "Climate change ground zero" owing to the marked changes in climate and geography, occasioned particularly by global warming (http://climatenexus.org/learn/regional-impacts/Alaska-climate-change-ground-zero). Temperatures are rising at twice the global rate. Alaska has warmed more than 2°C during the past 60 years, including an ~ 3°C increase in average winter temperature – a temperature change nearly half of the change that ended the last ice age. For example, the remarkably warm temperatures at the beginning of 2015 caused early snowmelt and the drying out of the Alaskan landscape, transforming areas of the normally wet and soft surface moss into a dry, brittle to fibrous layer.

This is a review of recent research on the changing environment and ecology of the vegetation, soil microbial communities, and the above- and belowground interactions of microbiota and vegetation in Alaska, with special attention to the tundra. Particular focus is placed on changes in vegetation revealed by satellite imagery and experimental ecological studies examining the effects of warming, soil drying, and changes in soil chemical and physical properties on vegetation and belowground microbial communities. Where appropriate relevant studies obtained from other locales are included to help elucidate findings from the tundra. Observational, real-time studies, of the changing tundra are also included, especially when they contribute to a clearer understanding of results from more indirect means such as satellite data and experimental field and laboratory-based studies. Given the main focus is on observational and experimental research, less attention is given to modeling studies, although this is a field of increasing importance as we endeavor to make better predictions about future scenarios.

There are four sections: Changing Arctic Vegetation, Soil Microbial Communities, Plant and Tundra soil Microbial Interactions, and Conclusions.

## CHANGING ARCTIC VEGETATION

Plants are among biota most sensitive to climate, including changing atmospheric and soil conditions. They are directly dependent for their survival and growth on atmospheric and terrestrial resources including, but not limited to, adequate moisture through precipitation and available moisture in the active soil layer, appropriate temperatures within the species tolerance range, sources of adequate light and concentrations of atmospheric carbon dioxide to support photosynthesis; and for vascular plants, sufficiently deep and nutrient rich soil to support strong and healthy root growth. Within the plant kingdom, tundra plant communities encompass a broad range of plant groups ranging from bryophytes (green moss) to vascular plants. The latter include soft-stemmed herbaceous plants such as sedges, grasses and some indigenous flowering plants, small shrubs such as birch (*Betula*), and small trees, including alder (*Alnus*) and willow (*Salix*). Conifers (evergreen trees) such as black and white spruce (*Picea*) are common in locales suitable for their growth. Lichens (fungal-algal symbioses) are also abundant and significant as sources of food for browsing animals such as caribou (*Rangifer tarandus*), especially in the Arctic during winter. Compared to other biomes, the biodiversity is low with only 1,700 species of vascular plants. The short growing season, and solid permafrost soil (frozen from 10 – 35 in. deep), that thaws only briefly for several months in summer to a shallow depth, prevents strong root development, thus historically limiting large tree growth. With global warming and more extensive thaw of the permafrost, there is a potential for greater incursion of larger vascular plants including taller trees as climate change progresses. Some current evidence of the shifting vegetation scene is presented in the following three sections: Satellite Data and Aerial Observation, Observational Field-based Research, and Experimental Field-based Research. Additional information is presented on the role of wildfires that are occurring increasingly in greater numbers, higher severity, and across a greater area, thus altering the vegetation, ecology, and soil properties of the Arctic.

## Satellite Data and Aerial Observations

The launching of earth-orbiting, data gathering satellites in the mid twentieth century (among them, the first U.S. satellite *Explorer 1*, 1958) opened a new era in Earth observatory capacity, with approximately 1,000 all totaled in operation by the second decade of the twentieth century. Remote sensing satellites are equipped with a variety of data-gathering sensors, including: Optical imaging systems (gathering visible, near infrared, and shortwave infrared data), Thermal imaging systems (for surface temperature), and Synthetic aperture radar (SAR) imaging systems (e.g., gathering vertical relief data). These systems provide major advantages in monitoring and assessing global vegetation patterns that are within the orbiting satellite's geographic data gathering range. The highest resolution instruments provide detail at dimensions of 5 m or less. Two major areas of vegetation analysis research using satellites are reviewed: Changes in vegetation over time, seasonally and across decades, and Vegetation biomass density and geographic coverage. Some emerging evidence of wildfire events and consequences is also reviewed. Naito and Cairns (2011) provide a useful overview of some recent advances in remote-sensing technology and its affordances in arctic vegetation dynamics relative to other major biomes, and their paper provides a potentially useful context for this section. They also make recommendations for future land management policies.

*Changes in vegetation over time, seasonally and across decades.* Satellite data on changes in arctic vegetation coupled, in some cases, with ground-based evidence has been particularly useful in investigating how vegetation patterns change seasonally and over decadal time scales. On a global basis, Eastman et al., (2013) analyzed a 30-year series (1982-

2011) of data from satellite spectral analyses of vegetation derived from the Global Inventory Modeling and Mapping Studies to detect the presence of trends in seasonality. Over half (56.30%) of land surfaces were found to exhibit significant trends, thus indicating that changes in seasonality are not a rare occurrence. Among these significant trends, almost half (46.10%) belonged to the following three classes of seasonal trends (or evidence of changes). The first significant class (Class 1) accounted for 20.43% of all significantly trending areas, while the second (Class 2) and third (Class 3) most frequent classes accounted for 16.54% and 9.13%, respectively. Class 1 consisted of areas that evidenced a uniform increase in vegetation greenness throughout the year, and was primarily associated with forested areas, particularly broadleaf forests. Class 2 consisted of areas that had evidence of an increase in amplitude of the annual seasonal signal. Thus increases in vegetation greenness in the growing season were balanced by decreases in the brown non-growing season. These areas were found primarily in grassland and shrubland regions. Class 3 was found primarily in the Tundra, as well as Taiga, biomes and exhibited increases in the annual summer peak in vegetation greenness. The results from this broad multi-decadal study for tundra, of special interest here, are consistent with other published observations. These include evidence of large declines in snow cover duration in Tundra Biomes during the 20th century in relation to increases in temperature, which extends the period of time for plant photosynthetic activity (Chapin III, et al., 2005; Euskirchen et al., 2007; Hollister et al., 2005). These observations suggest that seasonal environmental variables may be of particular significance in tundra phenology.

For example, early seasonal effects, such as time of snowmelt onset, in the Arctic can have profound effects on subsequent development of vegetation, depending on the geographic locale. Earlier snowmelt and longer growing seasons due to climate warming have been hypothesized to enhance vegetation productivity (e.g., Zeng et al., 2011). Gamon et al., (2013) used field observations and satellite data to examine vegetation phenology (seasonal cycles) and productivity patterns spatially and temporally for a coastal wet sedge tundra site near Barrow, AK during three growing seasons (2000-2002). Contrary to hypothesized predictions; earlier snowmelt did not lead to increased productivity. Higher elevated sites that became snow free earliest had relatively low evidence of productivity, whereas low-lying regions, that were slow to emerge from snow, reached the highest vegetation productivity by mid-season. Indeed, productivity was associated primarily with precipitation and soil moisture, and secondarily with growing degree days – the latter during this period yielding reduced growth in years with earlier snowmelt. Slight local depressions (typically 10–20 cm in depth) tended to have higher moisture content, often containing standing water for short intervals. These wet locations were dominated by vascular plants, particularly graminoids (e.g., grasses, sedges and rushes), whereas less productive higher, drier locations supported a higher percentage of lichens and mosses. These results point to the importance of combining fine scale and broader scale data gathering in refining predictions about the effects of earlier arctic growing seasons on vegetation productivity. However, the data of Gamon et al., (2013) support emerging evidence from field studies that early-season, local environmental conditions, especially moisture and temperature, are primary factors determining arctic vegetation productivity. For this northern coastal arctic site, growth conditions in the early growing season are mainly influenced by microtopography, hydrology, and regional sea ice dynamics. Predictions are less related to snow melt date or seasonal mean air temperatures, alone. Moreover, this study illustrates the merits of in-situ monitoring of the actual vegetation responses using field visual observations to obtain detailed information on surface conditions not possible from satellite observations, alone. More generally, studies such as this illustrate the importance of understanding regional variations in climate effects on vegetation dynamics.

As further evidence of regional and topographic effects on vegetation dynamics, Sweet et al., (2015) examined the possible effects of the ongoing increase of deciduous shrub abundance on plant canopy phenology and productivity during the growing and peak seasons in the arctic foothills region of Alaska. Using *in situ* spectral visual data, they compared deciduous shrub-dominated and evergeen/graminoid-dominated, community-level canopy phenology during the growing season. These studies used leaf area index (LAI) as an indicator of vegetation cover. The LAI is a measure of total leaf area of vegetation relative to the basal terrestrial area directly under the vegetation. They used a tundra plant-community-specific leaf area index (LAI) model to estimate LAI throughout the growing season and a tundra-specific net ecosystem carbon exchange (NEE) model to estimate the impact of increased deciduous shrub abundance and its associated shifts in both leaf area and canopy phenology on tundra carbon flux (exchange of $CO_2$ with the atmosphere). Deciduous shrub canopies reached onset of peak greenness 13 days earlier and the onset of senescence (die back) 3 days earlier compared to evergreen/graminoid canopies. This resulted in a 10-day extension of the peak season.

The combined effects of a longer peak season and the greater leaf area of deciduous shrub canopies almost tripled the modeled net photosynthetic carbon uptake of deciduous shrub communities compared to evergreen/graminoid communities. Moreover, the longer peak season alone resulted in 84% greater carbon uptake in deciduous shrub communities. These results suggest that greater deciduous shrub abundance increases carbon uptake not only due to greater leaf area, but also due to an extension of the period of peak greenness and productivity that prolongs the period of maximum carbon assimilation by photosynthetic fixation of $CO_2$.

This is consistent with a time series study reported by Potter (2014), based on satellite vegetation coverage data from 2000 to 2010, with the objective to understand landscape-level patterns of vegetation change in ecosystems of interior Alaska. The analyses of data-sets for Alaska vegetation cover types, wetland cover classes, wildfire boundaries since the 1940s, permafrost type, and elevation provided evidence to identify the most likely combination of factors driving regional changes in habitat quality and ecosystem productivity. Approximately 48% of all ecosystem geographic study areas in interior Alaska were identified with significant ($p < 0.05$) positive or negative growing season vegetation coverage trends from 2000 to 2010. Three-quarters of these ecosystem study areas (nearly 110,000 $km^2$) had evidence of significant positive growing season vegetation growth trends. The vast majority of interior Alaska areas with significant positive growing season vegetation growth trends were classified as upland shrub cover. However, in addition, non-forested wetlands (marshes, bogs, fens, and floodplains) were colocated with the shrublands on 13% of that total area. Overall, this evidence supports an hypothesis that temperature effects (warming) have markedly enhanced the rates of shrubland vegetation growth across interior Alaska in recent years.

In addition to seasonal studies, substantial evidence of the greening of the Arctic and expansion of shrublands has accumulated from multi-decadal studies of aerial photographs and satellite data (Bunn and Goetz, 2006; and as reviewed by Myers-Smith et al., 2011).

Satellite data for greening of arctic Alaska (1981-2001) was analyzed by Jia et al., (2003) for three bioclimate subzones in northern Alaska. They reported a 16.9% (±5.6%) increase in peak vegetation greenness across the region corresponding to simultaneous increases in temperatures. Changes in four specific vegetation types, using an 11-yr finer resolution satellite data, showed that the temporal changes in peak and time-integrated greenness were greatest in areas of moist nonacidic tundra. All totaled, these changes in greenness between 1981 and 2001 corresponded to approximately a 171 g $m^{-2}$ increase in aboveground plant biomass.

Sturm et al., (2001) using historic and modern aerial photographs concluded that during the past 50 years of warming, there has been a widespread increase in shrub abundance over more than 320 $km^2$ of the arctic landscape between the Brooks Range and arctic coast. They examined particularly changes in the three principal deciduous shrubs, dwarf birch (*Betula nana*), willow (*Salix* sp.) and green alder (*Alnus crispa*), as well as changes in treeline white spruce (*Picea glauca*) along the southern edge of the study at the northern base of the Brooks Range. The study area was in a location where human and natural disturbances (leading to successional changes) were minimal, so the authors attributed much of the increase in the abundance of shrubs to the recent change in climate. Moreover, the expanding incursion of vascular woody plants could be a significant contributor to changes in the high-latitude carbon budget, as well as contributing to important changes in the exchange of surface energy. Further studies by Tape et al., (2006) using data spanning 50 years have provided additional confirmation of these trends and provided more detailed vegetation cover changes at the local level. These results are consistent with other time series studies using satellite and aircraft aerial surveys of changes in vegetation in the Beringian Arctic (Bering Sea region). Lin et al., (2012) created multi-temporal high spatial resolution land cover maps for seven locations in he Beringian Arctic and assessed the change in land cover for the period of 1848-2008. Four of the five landscapes studied in Alaska underwent an expansion of drier land cover classes, while the two landscapes studied comparatively in Chukotka, Russia showed an expansion of wetter land cover types.

While there is broad consensus that the vegetation coverage and plant functional types are changing across the arctic landscape, regional differences are clearly to be expected and have been documented. Raynolds et al., (2013) examined 22 years of satellite data in an area of the Northern Foothills of the Brooks Range, analyzing an 823-$km^2$ area. They found that apparent homogeneous greening, revealed by coarse resolution satellite imagery, was very heterogeneous at finer resolution, with a strong influence due to glacial history. Small, scattered patches with significant increases in satellite evidence of expanded greening occurred throughout the younger, late Pleistocene glacial deposits. On older, mid-Pleistocene deposits, increases occurred in few, larger patches of mostly tussock-sedge, dwarf-shrub, and moss tundra. This is possibly a result of release of nutrients from thawing of ice-rich permafrost. Trends in evidence of greening varied by glacial history, elevation, slope, and the resulting vegetation conditions. The authors conclude that this heterogeneity in response to climate change can be expected throughout much of the Arctic, where complex glacial histories determine existing soil and vegetation characteristics. Further evidence of geographic heterogeneity was reported by Elmendorf et al., (2012) for 158 plant communities spread across 46 tundra locations. They found biome-wide trends of increased height of the plant canopy and maximum observed plant height for most vascular plant growth forms; an increased abundance of litter; increased abundance of evergreen, low-growing and tall shrubs;

and less evidence of bare ground. Although there was evidence of a correlation of increased vascular plant abundance and summer warming, the association was dependent on climate zone, moisture regime, and the presence of permafrost.

Pattison et al., (2015) examined local (plot-level) trends in species composition for field plots in the Arctic National Wildlife Refuge on the coastal plain in NE Alaska from 1984 to 2009 and linked these trends to trends in satellite evidence at fine and coarse scales. During this time, there were few changes in plant community composition. None of the five tundra types that were measured (1. wet sedge and sedge willow, 2. sedge-*Dryas*, 3. tussock and shrub, 4. riparian, and 5. *Dryas* terrace) had increases in total vegetative cover, and deciduous shrub cover in this coastal location exhibited none of the large increases reported elsewhere. Moreover, other studies at sites in northern Alaska (e.g., Tape et al., 2012), found that within tundra vegetation types, shrub expansion appears to be localized to higher resource environments such as floodplains, stream corridors, rocky outcrops, and/or in areas with deeper active layers. Similarly, Verbyla (2008) reported that satellite observations of trends in vegetation between 1982-2003 showed that cold arctic tundra significantly increased in vegetation coverage, while relatively warm and dry interior boreal forests areas consistently decreased. The annual maximum satellite evidence of greening in arctic tundra areas was strongly related to a summer warmth index as reported by Walker et al., (2003) in the next subsection on vegetation biomass.

*Vegetation biomass density and geographic coverage.* With increasing evidence of a likely expansion of woody shrubs and small trees in some parts of the Arctic, it is also important to study evidence of spatial and temporal patterns in the mass of vegetation (phytomass), plant height, and other ecologically significant metrics related to vegetation abundance and distribution. In this section, particular focus is placed on these functional morphological and anatomical aspects of tundra vegetation.

Detailed analyses of the effects of summer warmth on leaf area index (LAI), total aboveground phytomass (TAP), and density of vegetation greenness estimated by evidence from satellite imagery were made by Walker et al., (2003). Data were collected across the Arctic bioclimate zone in Alaska (extending from the northern boundary southward to the Brooks Range and along the west coast) and extrapolated to the entire circumpolar Arctic. The warmth for plant growth generally increases along the Alaskan bioclimate gradient from north to south. Alaskan phytomass, LAI, and greenness were related to the total summer warmth index (SWI) using statistical regression analyses based on data from 12 climate stations in northern Alaska. SWI = sum of mean monthly temperatures that exceed zero degrees Celsius (C). SWI varies from 9°C at Barrow on the northern edge of the Coastal Plain, AK to 37°C at Happy Valley on the southern edge of the Coastal Plain. Overall, based on the correlations, a five degrees C increase in the SWI is predicted to produce ~ a 120 g $m^{-2}$ increase in the aboveground phytomass for zonal vegetation on acidic soil sites, and ~ 60 g $m^{-2}$ on nonacidic sites. The distribution of acidic relative to nonacidic sites is related to the time since the last glaciation. Shrubs accounted for most of the increased vegetation on acidic substrates, whereas mosses accounted for most of the increase on nonacidic soils. The LAI was positively correlated with SWI on acidic sites, but not on nonacidic sites. The satellite evidence for green vegetation coverage was positively correlated with SWI on both acidic and nonacidic soils, but coverage on nonacidic parent material was consistently lower than the coverage on acidic substrates. Extrapolation to the entire Arctic using a five-subzonation approach to stratify the circumpolar satellite-derived vegetation coverage and phytomass estimates showed that 60% of the aboveground phytomass is concentrated in the low-shrub tundra, whereas the high northern Arctic has only 9% of the total. Estimated phytomass densities in five subzones from north to south are 47, 256, 102, 454, and 791 g $m^{-2}$, respectively. Based on the statistical evidence, climate warming is likely to result in increased phytomass, LAI, and green vegetation coverage on arctic localized geographic sites. The changes may be most noticeable in acidic areas with abundant shrub phytomass.

In addition to satellite visual evidence, Light Detection and Ranging (LiDAR) technology has recently been evaluated as a potential tool to estimate the tundra's vegetation composition and physical structure. LiDAR remote sensing - either from satellite, aircraft or on the ground - measures distance from sensor to a target object (e.g., a shrub) by emitting a laser beam that illuminates the target and analyzing the time it takes for the light to bounce off of the target and return back to its source (sensor location). Greaves et al., (2015) use ground-based LiDAR techniques to estimate biomass and leaf area of two dominant, low-stature (<1.5 m tall) arctic shrub species (*Betula* and *Salix*) in 24 (0.64 $m^2$) subplots established in northern Alaskan tundra (Toolik Lake). They found that LiDAR is a promising technology to monitor the biomass of woody plants such as woody shrubs and small trees that are increasingly invading previously moss-rich tundra, expanding on results of prior passive, spectral remote sensing techniques that were used to estimate dynamics in plant biomass (e.g., Boelman et al., 2003, 2005, 2011). For example, they found strong relationships between total harvested biomass and total leaf dry mass ($R^2 = 0.93$), and between leaf dry mass and leaf wet area ($R^2 = 0.99$). In this way, Greaves et al., show that their LiDAR based approach yields aboveground biomass and leaf area estimates for low-stature shrubs at fine spatial scales (sub-meter to ~50 meters) with the fidelity required to monitor small but ecologically meaningful changes in tundra structure. Importantly, this field-based study is the first to suggest that airborne LiDAR may indeed prove useful in detecting ongoing and expected changes in the tundra's 3-D vegetation structure. Greaves et al., results were comparable to results from previous studies linking TLS metrics to vegetation biomass in other biomes, including broad-leaved and conifer trees (Ku et al., 2012; Lin et al., 2010; Olsoy et al., 2014; Vierling et al., 2012). Beck et al., (2011) mapped shrub cover on the north slope of Alaska using satellite remote sensing images to distinguish between and map (>1m) and short shrub presence at a high spatial resolution (<5 m grid cell size) in different parts of the geographic region observed. The data were statistically transformed to yield maps of total- and tall-shrub cover, expressed as a percent of the total surface area. Data were also collected by direct observation of the coverage of the two groups of shrubs in the field, and the final shrub cover maps generated from their statistical analyses corresponded well with the field measurements ($r^2 = 0.7$). The resulting maps compared favorably with existing vegetation type maps previously published for the study area. More specifically, their shrub cover maps predicted that shrubs were present in 86% of the mapped area, with 78% of the area having a cover of ≥10%, and 54% having a cover of ≥50%. Overall, the maps showed that shrubs are nearly ubiquitously present on the North Slope of Alaska. They occurred in all but the wettest area, with low-lying wetland areas on the coastal plain of the North Slope having the lowest shrub cover. Tall shrub presence was predicted in 29% of the mapped area, with 18% having a cover of ≥10%, and 3% having a cover of ≥50%. For overall patterns, the map indicated a general east-west gradient of increasing shrub cover; tall shrub cover particularly was evident on the North Slope of Alaska. Moreover, further north on the North Slope, tall shrubs become more restricted to favorable topographical conditions such as stream channels and floodplains.

However, tall shrub cover over areas > 50% was rarely observed when total shrub cover was ≤60%. These data are consistent with the evidence of geographic variability reported in the final portions of the prior section on temporal changes in vegetation (e.g., Elmendorf et al., 2012; Pattison et al., 2015; Raynolds et al., 2013; Tape et al., 2012; Verbyla, 2008).

## Field-Based, Direct Observations

Satellite and other remote-sensing technology provide broad scale coverage of vegetation changes in the Alaskan tundra with varying detail depending on fine-scale resolution, but detailed studies of plant morphology, physiology and other biological aspects of importance in vegetation dynamics at site-levels are currently best studied by direct data gathering in the field. To augment the review of remote-sensing studies presented above, some illustrative field-based studies are included here. Some general published sources can be consulted as a context for this section (e.g., Callahan and Jonasson, 1995; Chapin III et al., 2005; Epstein et al., 2004; Hufnagel and Garamvölgyi, 2014; Post et al., 2009; Walter, 2004).

Seasonal studies using direct observation at various sites in Alaska have provided some general insights into climatic forcing functions on vegetation composition and diversity. For example, observations of vascular plants, bryophytes and lichens in sub arctic tundra by Makoto and Klaminder (2012) indicated that soil frost disturbances are important for maintaining successional gradients over several centuries within the arctic landscape at small spatial scales (<3m). Moreover, they predict that the termination of soil frost activity as a result of future warmer winters may result in a loss of micro-sites having young vegetation communities with high plant diversities, and a subsequent establishment of mature shrub-dominated plant communities. However, Frost et al., (2013) show that increasing shrub dominance is not occurring as a simple function of regional climate trends, but is also a function of cryogenic disturbances associated with small, widely spread patterned-ground landscapes throughout the Arctic. Based on evidence that temperature seasonality (S-T), i.e., the difference between summer and winter temperatures, is diminishing over time with warming of northern lands, Xu et al., (2013) using a combination of remote sensing and land-based data report that trends in the timing of initiation, termination and performance of photosynthetic activity, tied to threshold temperatures, may alter vegetation productivity or modify vegetation seasonality (S-V) over time. The observed diminishment of S-T and S-V indicates a significant latitudinal shift equatorward, from more northern to southern limits, during the past 30 years in the Arctic. With continued warming, they predict that an additional S-T diminishment equivalent to a $20^{o}$ equatorward shift could occur this century, and more attention is needed to determine the impact that this may have on the environment and ecosystem services.

Seasonal warming also can have major affects on topography and localized soil conditions. Continued warming trends affect plant growth directly, but also can have indirect affects through changes in nutrient availability and soil structure. Warming can cause permafrost to thaw and thermokarst (ground subsidence) to develop. This can alter the structure of the ecosystem by altering hydrological patterns within a site and driving changes in the composition of vegetation (e.g., Osterkamp et al., 2009). Schuur et al., (2007) examined a natural gradient of permafrost thawing in three Alaskan tundra sites. They found that vascular plant biomass shifted from graminoid-dominated tundra in the least disturbed site to shrub-dominated tundra at the oldest, most subsided site. The intermediate site, however, was co-dominated by both plant functional groups. Patterns of productivity for vascular plants followed the changes in biomass, whereas nonvascular moss productivity was especially important in the oldest, most subsided site. Across all sties, graminoids were most evident on the cold, dry microsites; whereas, the moss and shrubs were associated with the warm, moist microsites. Total nitrogen within green plant biomass differed across sites, suggesting that there were increases in soil nitrogen availability where permafrost had thawed.

Arndal et al., (2009) studied seasonal variation in gross ecosystem production, plant biomass, and carbon and nitrogen pools in five high arctic vegetation types (*Cassiope*, *Dryas*, *Salix* heath, grassland, and fell). Large differences were observed in seasonal growth and production within and among vegetation types. Mosses contributed considerably to the total C and N pool in grassland, fen, and *Salix* heath. The fell, with highest pool of leaf N, leaf chlorophyll, and moss N, was the most productive vegetation type in terms of gross ecosystem production (GEP), despite the lowest total biomass. Across vegetation types, leaf biomass, leaf N, and moss N pool size substantially influenced GEP. Within most vegetation types, GEP correlated with leaf N. This is correspondent with the notion that N may limit plant production in many high arctic ecosystems. The timing of the peaks in C and N pools in leaves did not coincide with that in the mosses and in woody tissues. With continued warming and shifts in plant functional types across arctic sites, there also may be unpredictable changes in the nutrient dynamics and plant productivity as exemplified by this research.

Shifting vegetation patterns also have a ripple effect throughout arctic ecosystems affecting wildlife (e.g., Marcot et al., 2015), including birds (Boelman et al., 2014; Fossøy et al., 2014; van Oudenhove et al., 2014; Wild et al., 2015), and other animals important as human food sources, such as caribou and other large and small mammals (Gustine et al., 2014; Joly et al., 2012; Nicolson et al., 2013; Wheeler et al., 2015; Vors and Boyce, 2009).

## Field-Based, Experimental Studies

Field-based observational studies have provided a sound framework for more detailed experimental studies on climate change in arctic tundra ecosystems. Some long-term experimental field research stations have been established in the Arctic, especially in the north coastal plain, including the International Tundra Experiment (ITEX), and the Arctic LTER suite of experiments near Toolik Lake, Alaska. Some of the major findings from these experimental studies are summarized here.

Henry and Molau (1997) and Oberbauer et al., (2013) summarized some of the major findings from the ITEX studies. The ITEX research used open-top chambers (OTCs), consisting of enclosing sides with an open top that permits ambient insolation, while also controlling temperature and regulating snow depth. Some of the major findings for vascular plants are presented. The OTC treatments increased mean near-surface temperatures by 1-3°C during the growing season, simulating predictions from global circulation models. All of the species investigated responded to the temperature increase, especially in phenology and reproductive variables. However, these short-term responses were differentiated, and no general pattern in type or magnitude of response was noted for functional types or phenology class. In general, responses were similar among sites, although the response magnitude tended to be greater in high arctic sites. Early snowmelt increased carbon:nutrient ratios in plants, suggesting that sustained growth and reproductive responses to warming will depend on nutrient supply. Moreover, increased carbon:nutrient ratios in litter could buffer nutrient cycling, and hence plant growth.

To obtain baseline data on natural variations in tundra climate on vascular plant responses, ITEX control data obtained from natural sites outside of the OTCs were used to test the phenological responses to background temperature variation across sites spanning latitudinal and moisture gradients. Overall, the analyses did not show an advance in phenology. Instead, temperature variability during the years sampled and an absence of warming at some sites resulted in mixed responses. Seasonal, phenological transitions of high arctic plants clearly occurred at lower heat sum thresholds than those of low arctic and alpine plants. Heat sum threshold is the initial temperature value required to sustain a plant response during the remaining season, such as leaf flushing in spring. However, sensitivity to temperature change was similar among plants from the different climate zones. Plants of different communities and growth forms differed for some phenological responses. Heat sums associated with flowering and greening appear to have increased over time. These results point to a complex suite of changes in plant communities and ecosystem functions in high latitudes and elevations as the climate warms, and also points to some of the challenges of adequately simulating natural environmental changes in experimental studies.

Borner et al., (2008) reported more detailed data on ITEX experiments where the effects of increased winter snow depth, and thus decreased growing season length, were examined on the phenology of four arctic plant species (*Betula nana*, *Salix pulchra*, *Eriophorum vaginatum*, and *Vaccinium vitis-idaea*), including seasonal nitrogen availability in arctic snowbed communities. Increased snow depth had a major effect on the temporal pattern of first snow-free date in spring, and plant bud break, and flowering; but did not affect the rate of plant development. By contrast, snow depth had a large qualitative effect on N mineralization in zones with deep snow, causing a shift in the timing and amount of N mineralized compared to the control ambient snow zones. Nitrogen mineralization in deep snow zones occurred mainly overwinter; whereas, N mineralization in ambient snow zones occurred mainly in spring. Concentrations of soil dissolved organic nitrogen (DON) were approximately five times greater than concentrations of inorganic nitrogen (DIN) and did not vary significantly over the season. Based on these results, the authors predicted that increases in the depth and duration of snow cover in arctic plant communities will likely have minor effects on the rate of plant phenological development, but potentially large effects on patterns of N cycling.

Rumpf et al., (2014) further examined the effect of snow regimes (snow cover depth and duration) by experimentally manipulating snow regimes using snow fences and shoveling. They assessed above ground size of eight common high arctic plant species weekly throughout the summer. Plant growth responded to snow regime. Air temperature sum during the snow free period was the best predictor for plant size. Plants in early snow-free treatments without additional spring warming were smaller than controls. Responses varied among species to the effects of deeper snow with later melt-out. Moreover, no generic trends were detected when responses were categorized by growth forms or habitat associations, thus indicating the importance of examining species-level responses to some of these snow regime variables.

Using a combined approach of snow fencing and experimental warming in OTCs Wahren et al., (2005) examined the effect of changes in winter snow cover at Toolik Lake, Alaska. OTCs, paired with unwarmed plots, were placed along snow gradients for each experimental and control areas that were adjacent to snowdrifts. After eight years, the vegetation of the two sites, including that in the control plots had changed significantly. At both sites, the cover of shrubs, live vegetation, and litter, together with canopy height, had all increased; while lichen cover and diversity had decreased. At the moist site, bryophytes decreased in cover, while an increase in graminoids was almost entirely due to the response of the cottongrass sedge *Eriophorum vaginatum*. The snow addition treatment particularly affected species abundance, canopy height, and diversity; whereas, the summer warming treatment had few measurable effects on vegetation. However, the latter must be interpreted in the context that the natural interannual temperature fluctuation at these sites was considerably larger than the temperature increases within OTCs (< 2 °C). Snow addition also had a greater effect on microclimate by insulating vegetation from winter wind and temperature extremes, modifying winter soil temperatures, and increasing spring run-off. Most increases in shrub cover and canopy height occurred in the medium snow-depth zone (0.5–2 m) of the moist site, and the medium to deep snow-depth zone (2–3 m) of the dry site. Deciduous shrubs, particularly *Betula nana*, increased in cover at the moist tundra site, while evergreen shrubs decreased. These differential responses may be explained, in part, by the larger production to biomass ratio in deciduous shrubs, combined with their more flexible growth response under changing environmental conditions. At the dry site, where deciduous shrubs were a minor part of the vegetation, evergreen shrubs increased in both cover and canopy height.

On a broader scale, Van Wijk et al., (2003) used meta-analysis to examine the results of the Arctic LTER experiments near Toolik Lake, Alaska, including some comparative data from Abisko, Sweden; with special emphasis on aboveground biomass responses of different arctic and subarctic ecosystems to experimental fertilization, warming and shading. The results for Toolik Lake are reviewed here. While there were some consistent trends, site-specific differences were noted as reported in other tundra experimental research. Aboveground plant biomass, particularly the biomass of deciduous and graminoid plants, responded most strongly to nutrient addition. In contrast, evergreen shrubs showed a significant negative overall response to fertilization, although this was caused entirely by a strong decline of the biomass. The biomass of mosses and lichens decreased as the biomass of vascular plants increased. The decreased response of the non-vascular species is probably caused by an inhibition in growth by a combination of shading from the dense upper canopy of *Betula nana* and burial by vascular plant litter. As reported in other tundra warming studies, *Betula nana* increased its dominance and replaced many of the other plant types. The warming without fertilizer addition did not lead to any significant responses among the different vascular plant types, and shading did not lead to significant effects in any group of the vascular plants. In general, these results are consistent with other experimental studies, where fertilization has a greater effect on vascular plant biomass than warming. Moreover, there were also large site-specific differences within each region. The variations in response patterns show the need for analyses of joint data sets from many regions and sites, in order to uncover common responses to changes in climate across large arctic regions in comparison to regional or local responses.

Some additional results of experimental studies of vegetation responses to changing climate variables include the following: herbaceous plant and shrub communities (Bret-Harte et al., 2008; Chapin III et al.,1995; Gough and Hobbie 2003; Heskel et al., 2014; Jagerbrand et al., 2012; Klady et al., 2011; Lang et al., 2012; Mack et al., 2004; Marchand et al., 2005; McLaughlin et al., 2014; Shaver et al., 2001), and trees (Hobbie and Chapin, 1998; Hofgaard et al., 2010; Moyes et al., 2013).

## Fire Consequences

Some illustrative studies on fire effects in the tundra are reviewed. While there is good evidence of increasing invasion of woody vascular plants in susceptible regions of the Arctic; wildfires are reported to be more frequent, increasingly severe, and of larger spatial extent due in part to warming and drier conditions. However, fire effects on tundra ecosystems are poorly understood and sometimes difficult to quantify in remote regions where a short growing season seriously limits ground data collection. Kolden and Rogan (2013) used satellite coarse-resolution remote sensing to quantify wildfire burn severity of the 2007 Anaktuvuk River Fire in Alaska, the largest tundra wildfire that has been recorded on Alaska's North Slope. The satellite data were processed to provide broad scale evidence of surface, subsurface, and comprehensive burn severity. They analyzed the burn relative to three temporal periods: Pre-fire period, Initial Assessment post-fire (pre-green up), and Extended assessment post-fire (post-green up). The pre-fire situation consisted of low-lying herbaceous material (including grass tussocks) on less productive sites, or continuous dwarf shrub canopy normally <1 m in height on more productive sites. Bare soil was not a component of the pre-fire scene in this ecotype, as an organic horizon of decomposing biomass covered the mineral soil. As a result of the burn, the landscape was marked by localized zones with standing water, burned and saturated char and black soil surfaces, and patches (sub-meter to several meters in area) of unburned or only partially consumed non-photosynthetic vegetation, consisting primarily of grasses from prior growing seasons. Moreover, the organic horizon was consumed in a spatially variable pattern up to a meter in depth. At one year post-burn, data showed a decrease in severity, including added photosynthetic vegetation. Tussock sedges regenerated during summer in some of the most severely burned sites, in general indicating rapid vegetation regeneration on the burned site at this locale.

Overall, the Anaktuvuk River fire burned 1,039 square kilometers of the Alaskan Arctic Slope with serious consequences for changes in the soil and ecosystem carbon content (Mack et al., 2011). The tundra ecosystems lost 2,016 ± 435 g carbon $m^{-2}$ in the fire, an amount two orders of magnitude larger than the annual net carbon exchange through photosynthesis by plants and $CO_2$ emissions due to respiration and other sources in undisturbed tundra. Sixty percent of this carbon loss was from soil organic matter, and radiocarbon dating of residual soil layers revealed that the maximum age of soil carbon lost was 50 years. In proportion to the entire burned area, the fire released approximately 2.1 teragrams of C to the atmosphere, an amount similar in magnitude to the annual net carbon sink for the entire arctic tundra biome averaged over the last quarter of the twentieth century. Further documentation of the Anaktuvuk River fire behavior and burn severity is provided by Jones et al., (2009).

The persistence of post-burn, long-term effects following tundra wildfires was reported by Barrett et al., (2012) at a time 17 years after a tundra fire on the North Slope of Alaska. Fire-related changes in vegetation composition were assessed from remote-sensing imagery and ground observations of the burn scar and an adjacent comparative control site. Early-season remotely sensed imagery from the burn scar exhibited a low vegetation coverage compared with the control site. However, the late-season evidence is slightly higher. The satellite data of vegetation coverage indicated a quick recovery, reaching the range of pre-fire levels three years after the burn, with occasional spikes that were much higher. After five years of recovery, the maximum growing season evidence of vegetation greening was elevated in the burn scar although there was no difference in the average extent compared with pre-fire levels. The burned sites had 86% and 91% ground cover, whereas the control site had 100% vegetation cover. The ground cover composition in the burned sites was distinct from that of the control. The latter was composed primarily of *Salix*, leaf litter and, to a lesser extent, *Eriophorum vaginatum* (cottongrass) and other graminoids. Although *E. vaginatum*, *Salix* spp. and litter were major components of the burned site, moss and *Betula nana* were considerably less abundant. In addition to grasses, ground cover typically found in the burned site, but not the control, included *Ledum palustre* (an evergreen shrub in the heath family: Ericaceae), forbs, fireweed (*Chamerion angustifolium*) and open ground.

In a similar designed study, Narita et al., (2015) reported on the recovery of tundra vegetation and the depth of permafrost thaw on the Seward Peninsula, Alaska, the site of a wildfire in 2002. As in the study by Barrett et al., vegetation in the burned site was compared to an adjacent unburned tundra site five to 10 years post-fire. Effects of the fire on the vegetation varied among species and were spatially variable at the stand scale. Notably, the cover of evergreen shrubs, bryophytes, and lichens remained drastically decreased five years after the fire and had not recovered even 10 years after the fire. However, the cover of graminoids, especially *E. vaginatum*, and of the deciduous shrub *Vaccinium uliginosum* increased. The depth of permafrost thaw increased, and its spatial pattern was related to vegetation structure; specifically, deeper thaw corresponded to graminoid-rich areas, and shallower thaw corresponded to shrub-rich areas. As the *E. vaginatum* cover increased, the thaw depth recovered to that of the unburned area, and the spatial variation had disappeared 10 years after the fire. Further evidence of physical short-term changes after a tussock tundra fire on the Seward Peninsula is presented by Liljedahl et al., (2007).

Further chronological evidence of burn recovery was documented by Racine et al., (2004) who also studied a tundra fire and vegetation change on the Seward Peninsula, Alaska, but in this case along a hillslope where the recovery was not so pronounced as at the site studied by Narita et al., In this study, prior to the fire in 1977, soils and vegetation ranged from poorly drained moist tussock-shrub tundra on the lower slopes to well-drained dwarf shrub tundra on the back slope and very poorly drained wet sedge meadow on the flat crest. The vegetation was sampled on the slope before the fire and at eight sites following the fire at irregular intervals from one year to 25 years. Short-term recovery, during the first decade, was dominated by growth of bryophytes, sedges, and grasses from both regrowing sedge tussocks and seedlings. However, during the second and third decade, and by 24 years after the fire, evergreen (*Ledum palustre*) and deciduous shrubs (mainly *Salix pulchra,* willow) expanded markedly. Indeed, the shrub cover was generally higher than before the fire! Upslope on the better-drained and more severely burned tussock-shrub and dwarf-shrub tundra sites, willows (mainly growing from seed) became established during the first 10 years after the fire, and appeared to have grown rapidly during the subsequent 15 to 20 years. However, *Sphagnum* moss and fruticose lichens showed little or no recovery after 24 years at any site, except for Sphagnum moss in the wet meadow site. Additional evidence of slow recovery of lichens on burned caribou range in the Alaska tundra has been reported by Jandt et al., (2008), including evidence of poor recovery after as much as 25 years post burn. Further evidence of the severity and consequences of Alaskan wildfires are reported by Rocha and Shaver (2011), Bret-Harte et al., (2015), and Loboda et al., (2013).

Beyond the tundra sites reviewed here, increasing evidence of the effects of wildfires on tree stands and woody plants has also been reported for a variety of forested sites in Alaska (e.g., Kaischke et al., 2002; Lloyd et al., 2007; Yarie, 1981).

## Plant and Tundra Soil Microbial Communities

Plants have a dynamic relationship with soil microbial communities (bacteria, fungi and protists). The microbes, especially in close vicinity of plant roots, increase the fertility of the soil by making essential plant nutrients more accessible through remineralization (Adl, 2003; Clarholm, 1981, 1989; Koller et al., 2013). Likewise, plant roots secrete organic nutrients utilized by fungi and bacteria for growth, and in turn the bacteria at the base of soil microbial foodwebs provide prey for protists, such as amoebae and flagellates (Adl, 2003; Anderson and McGuire, 2013; Darbyshire, 1994). Some larger amoebae also prey on fungi (Old and Darbyshire, 1978, 1980). Climate change, especially global warming, is expected to have major effects on the composition and life activities of tundra soil microbes. Some recent evidence is presented first, followed by implications for plant-microbe interactions under conditions of warming in the tundra.

### Global Warming Effects on Tundra Soil Microbial Communities

*Bacteria.* One of the most immediate effects of global warming is permafrost thaw and its consequences on microbial community structure and function, especially for bacteria. Deng et al., (2015) investigated soil bacterial and archael communities using molecular genetic techniques across a permafrost thaw gradient at different depths in Alaska during a thaw progression over three decades. Based on 97 samples, corresponding to 61 known classes and 470 genera, they reported that soil thaw depth and the associated soil physical-chemical properties had predominant impacts on the diversity and composition of the microbial communities. Both richness and evenness of taxa in the microbial communities decreased with soil depth. Acidobacteria, Verrucomicrobia, Alpha- and Gamma-Proteobacteria dominated the microbial communities in the upper horizon; whereas, abundances of Bacteroidetes, Delta-Proteobacteria and Firmicutes increased towards deeper soils. There were less effects of thaw progression in microbial communities in the near-surface organic soil, probably due to greater temperature variation. Thaw progression decreased the abundances of the majority of the associated taxa in the lower organic soil, but increased the abundances of those in the mineral soil, including groups potentially involved in degradation of recalcitrant C compounds (Actinomycetales, Chitinophaga, etc.). The changes in microbial communities may be related to altered soil carbon sources by thaw progression. Overall, this study revealed different impacts of thaw in the organic and mineral horizons and suggests the importance of studying both the upper and deeper soils while evaluating microbial responses to permafrost thaw.

Further evidence of impacts of changing soil physical-chemical properties on bacterial communities was studied by Campbell et al., (2010), especially with respect to the likely influence of increased deposition of reactive nitrogen that accompanies changing patterns of precipitation and permafrost thaw. Substantial losses of C were previously reported after long-term nutrient additions in moist acidic tundra (MAT) soils on the North Slope of the Brooks Range, Alaska, possibly due to the enhanced respiratory activity of more active bacteria (Mack et al., 2004). To assess the possible role of bacterial communities in these C losses, Campbell et al., utilized molecular genetic analyses coupled with community-level physiological profiling to describe changes in MAT bacterial communities after short- and long-term nutrient fertilization. They analyzed soil from fours sets of paired control and fertilized MAT soil sites. Bacterial diversity was lower in long-term fertilized plots. Long-term fertilization also was correlated with shifts in the utilization of specific substrates by microbes present in the soils. The combined data indicate that long-term fertilization produced a significant change in microbial community structure and function. This was linked to changes in C and N availability and shifts in aboveground plant communities.

Additional evidence of increased carbon loss from humic soil substances (HS) due to bacterial action in subarctic tundra soil (Council, AK) was reported by Park et al., (2015) using experimental microcosms. The quantity of humic acids (HA) decreased to 48% after a 99-day incubation at 5°C as part of a biologically mediated process. Bacterial community analysis showed that during the microcosm experiments, the relative abundance of bacteria and archaea (methane-producing microbes: "methanogens") particularly increased, suggesting their involvement in HS degradation. Furthermore, using 122 strains of HA-degrading bacteria cultured from nearby sites, the authors reported increasing HS-degradation rates in parallel with rising temperatures in a range of 0°C to 20°C, with most notable increase occurring at 8°C compared to 6°C. Overall, the results indicated that, although microbial-mediated HS degradation occurred at temperature as low as 5°C in tundra ecosystems, increasing soil temperature caused by global climate change could substantially increase HS degradation rates. Furthermore, if the thawing period is extended, degradation activity could also increase, thereby directly affecting nearby microbial communities and the plant rhizosphere environments surrounding plant roots.

Effects of global warming in the Arctic also have been studied with respect to the combined responses of bacteria and fungi (e.g., Wallenstein et al., 2007). Fungal and bacterial community structure in tussock, inter-tussock and shrub organic and mineral soils at Toolik Lake, AK were evaluated using molecular genetic techniques. The soil communities were sampled and analyzed at the end of the growing season in August 2004 and also just after the soils thawed in June 2005. Although tussock and inter-tussock soil communities were very similar at the phyla level, the communities differed substantially between vegetation types. Across sampling dates, the communities were relatively stable at the phyla and subphyla levels, but differed significantly at finer phylogenetic scales. Acidobacteria dominated tussock and inter-tussock bacterial communities, while Proteobacteria dominated shrub soils. This is consistent with previous reports that shrub soils contain an active, bioavailable C fraction, while tussock soils are dominated by more recalcitrant substrates. Concurrently, tussock fungi communities had higher proportions of Ascomycota than shrub soils, while Zygomycota were more abundant in shrub soils. Increasing evidence of greater shrub abundance in warmer and drier arctic sites suggests that soil microbial communities and their functioning are likely to be altered by continued climate change.

*Fungi.* Specifically, with respect to fungi in the Arctic, Timling and Taylor (2012) reviewed recent molecular data and concluded that there is comparatively high fungal diversity in arctic soils, with simultaneously no evidence for lower species richness at higher latitudes. Moreover, laboratory analyses of tundra soil fungal C-biomass indicates it is approximately an order of magnitude greater than bacteria, the next most abundant microbial taxa in most soils. For example, Anderson and McGuire (2013) reported tundra soil fungal C-biomass of 5 to 11 mg $g^{-1}$ soil dry weight compared to 20 to 120 μg $g^{-1}$ for total bacteria. The dominant fungi, and particularly ectomycorrhizal-forming fungi (ECM), appear to be cosmopolitan species. More particularly, community composition is altered under experimental warming, although arctic fungi are capable of growth at sub-zero temperatures. Melanized forms are frequent and host specificity is low. Experimental research on the effects of soil warming on tundra fungal communities (e.g., Gemi et al., 2015; Morgado et al., 2015; Semenova et al., 2015) has indicated that certain species of ECM and other fungal groups are favored by warming and may become more abundant, while many other species may go locally extinct due to direct or indirect effects of warming. Differences are also observed in moist tundra compared to dry tundra. There is a greater change in community composition in moist tundra, but less in the dry tundra. In general, species-level differences in responses to warming need to be considered more carefully. Moreover, such shifts in fungal community composition may also affect nutrient cycling and soil organic C storage as reported above for bacteria.

Moreover, other long-term warming studies of ECM associated with *Betula* shrubs (Deslippe et al., 2011) demonstrated opposing effects of long-term warming and fertilization treatments on ECM fungal diversity. Warming increased, and fertilization decreased, the diversity of ECM communities. They showed that warming leads to a significant increase in high biomass fungi with proteolytic capacity, especially *Cortinarius* spp., and a reduction of fungi with high affinities for labile N, especially *Russula* spp. In contrast, fertilization treatments led to relatively small changes in the composition of the ECM community, but increased the abundance of saprotrophs. Consequently, these data suggest that warming profoundly alters nutrient cycling in tundra, and may facilitate the expansion of *B. nana* through the formation of mycorrhizal networks of larger size.

*Protists (e.g., amoebae and flagellates).* Among the major bacterial predators in soil ecosystems, naked amoebae and heterotrophic flagellates are the most abundant, including those in tundra soils (e.g., Anderson 2014a). Given the importance of the soil microbial communities in ecosystem services, especially the critical link of protists between bacteria and higher biota in soil food webs, there is remarkably little current research on global warming and its effects on tundra soil protists. Indeed, much of the present literature is largely very recent (e.g., Anderson 2008, 2010, 2012, 2014b) and some of it is summarized here. In general, the densities of bacteria and protists in moss-rich tundra soils are comparable to organically rich soils elsewhere; e.g., bacteria ($10^8$-$10^9$ $g^{-1}$soil dry wt.), heterotrophic nanoflagellates ($10^5$-$10^7$ $g^{-1}$soil dry wt.), naked amoebae ($10^3$ $g^{-1}$soil dry wt.), and testate amoebae enclosed by an organic or mineral test (~$10^2$ $g^{-1}$soil dry wt.). In general, the higher the level in the trophic hierarchy, the larger the size of the protist, with some testate amoebae near the top in the size range of hundreds of microns.

Presently, we have less information on how global warming will affect the taxonomic composition and community structure of tundra soil protists compared to research on their physiological ecology, and role in the biogeochemical carbon cycle. The latter is of critical importance, because the amount of respiratory $CO_2$ released by soil microbial communities can be substantial, thus increasing atmospheric $CO_2$ and exacerbating the greenhouse effect. Also, bacterivorous protists can sequester carbon consumed from bacteria during predation, thus passing it up the food chain, and controlling bacterial biomass and bacterial respiration, a major source of soil microbial respiratory $CO_2$. Overall, based on laboratory studies of respiratory $CO_2$ release with increasing temperature, the $Q_{10}$ (rate of increase in respiration for each 10 deg. C rise in temperature) on average is very close to 2.0. That is, the metabolic activity of the soil microbes, in general, tends to increase by two times, for each 10 degree Celsius rise in temperature. Predictions of total $CO_2$ emissions from soil microbiota is expected to increase as global warming causes increased warming and thawing of the permafrost soil in the Arctic, resulting in release of organic compounds that can be used by the microbiota for respiration.

One model, based on laboratory evidence (Anderson, 2008, 2010), is that at 20°C, (approximate high summer temperatures in the tundra) the amount of $CO_2$ emitted to the atmosphere could increase from about 1 Kmol $CO_2$ $km^{-2}$ soil $h^{-1}$ to more than 3 Kmol $km^{-2}$ soil $h^{-1}$ during early spring conditions when the soil thaws from 5 cm depth to 15 cm depth. During summer when the microbial communities are more abundant, the emissions could be ~ four times as great. As reported in the above section on bacteria, altered C sources in the soil may significantly affect microbial community structure and function, including changing nutrient patterns, thus affecting both microbes and plants that depend on the nutrients. Moreover, the fate of soil-released respiratory $CO_2$ is of critical importance. If aboveground increase in plant production is sufficient, it may partially or completely offset the release from soil by fixing the $CO_2$ into organic compounds and plant biomass during photosynthesis (e.g., Anderson, 2013; Chapin and Shaver, 1996; Heskel et al., 2013; Sweet et al., 2015).

*Above- and belowground interactions of plants and soil microbial communities.* With increasing evidence of the expansion of herbaceous plants and especially shrubs and small trees into the arctic tundra, major changes can be expected in the dynamics of the interaction of the aboveground vegetation with the belowground microbial communities, especially the effect of roots on the soil physical structure and chemical composition. Increasing atmospheric $CO_2$ concentrations and increased warming, resulting in enhanced plant photosynthesis, also can elevate organic compounds translocated to the roots where some are released as root exudates. Root exudates include glucose, amino acids, tricarboxylic acids, and other small molecular weight organic compounds in addition to larger molecular weight products such as mucopolysaccharides and other gelatinous secretions, especially at the root tip. Organic root exudates, especially in the very thin soil layer surrounding the roots (rhizosphere), serve as organic nutrients that enhance bacterial growth and production. In turn, protists grazing on the bacteria proliferate in the $CO_2$-driven, organically enriched, soil environment (e.g., Anderson and Griffin, 2001; Rønn et al., 2002; 2003; Treonis and Lussenhop, 1997). Leaf fall and litter also contribute to organic enrichment of the soil and may particularly provide particulate organic C resources for fungi and other decomposers (Adl, 2003). Increased densities of microbial communities in organically enriched soil are expected to produce higher emissions of respiratory $CO_2$, thus potentially contributing to higher atmospheric loading of $CO_2$. Current laboratory-based studies on the effects of soluble organic compounds on tundra soil microbial respiration indicate that pulsed release of small molecular weight compounds into the soil produce a transient spike in respiration lasting on the order of an hour or more with increased $CO_2$ release (Anderson, 2012).

For example, glucose enrichment of tundra soil produced a pronounced two- to three-fold increase in respiration above basal rate, which declined over four hours to baseline levels. However, less than 1% (w/w) of glucose-C supplement was respired during the respiratory spike. A more substantial amount of the glucose-C became incorporated in microbial biomass. Although respiratory response to pulsed glucose-C was minimal, the overall mean basal rate at 20°C after one week ranged between 4 and 6 nmol $min^{-1}$ $g^{-1}$ soil, indicating a significant assimilation and respiration of constituent soil organic C. Respiratory $CO_2$ emissions were greater in summer than in early spring soil samples, probably related to more robust densities of microbiota that develop during the warmer summer months. Further research is needed on the synergistic or antagonistic effects of changing climate and edaphic variables among tundra environments, including permafrost thawing and nutrient release, plant responses to global warming, and the responses of soil microbes in a changing tundra climate. Some further published sources on issues of above- and belowground interactions in high latitude environments may be of interest (Bardgett and Wardle, 2010; Gough et al., 2012; Grogan and Chapin, 2000; Steiglitz et al., 2006).

## Conclusion

Together, field-based and remote-sensing studies have decisively documented that the arctic tundra vegetation is changing. With warmer temperatures, longer growing seasons, and in some cases drier conditions, more productive, woody vegetation is becoming more dominant on the tundra, at the expense of formerly moss-rich and graminoid-dominated tundra environments. These changes are also bringing less favorable conditions for lichen growth and herbaceous plants. This is due especially to shading by the canopies of the woody plants, cover by litter fall, and drier conditions; thus reducing the lichen and herbaceous plant biomass in many areas of the tundra. Herbaceous plants and lichens are a major source of food for caribou, the lichens are especially important in winter, and diminishing supplies may impact the health, geographic distribution and population of caribou. Woody plants also may be impediments for some ungulates in reaching foraging sites. Bird species, especially migratory birds that reproduce in the Arctic and smaller mammals (e.g., voles, lemmings and ground squirrels), are increasingly adversely affected by the changing environment. Amidst this general scenario, however, there are locales where these major changes are less evident, and the topography, hydrographics, and nutrient regimes of varying tundra locales contribute substantially to local differences in the vegetation response to a changing climate.

Changes in the aboveground vegetation also can produce favorable, and in some cases unfavorable, effects on belowground microbial communities, including bacteria, fungi and protists. Increased herbaceous and woody vegetation can provide more organic nutrients for microbial communities in addition to the release of organic compounds from thawing permafrost soil. Warmer growing conditions and higher concentrations of atmospheric $CO_2$ enhance plant productivity and $CO_2$-fixing photosynthesis, providing more organic compounds that are translocated to the plant roots. Organic exudates from the roots provide nourishment for bacteria and fungi. Bacteria are prey for protists (amoebae and small flagellates) at the base of soil food webs, and this can enhance the transfer of carbon compounds to higher trophic levels in the food web hierarchy. However, increased metabolism by soil microbes also yields higher emissions of respiratory $CO_2$ from soil to the atmosphere, and unless there is a compensatory uptake of the $CO_2$ by the increasingly denser vegetation, the net excess of atmospheric $CO_2$ may further exacerbate the greenhouse effect and global warming, particularly given the massive area of the circumpolar tundra. In some cases, permafrost thaw and increased precipitation leads to water logging of the moss-rich soil. While this potentially provides a favorable environment for growth of some soil protists, water logging can also lead to anaerobic conditions that suppress metabolism and prevent aerobic respiration resulting in less $CO_2$ emissions. However, anaerobic conditions favor the proliferation of methanogenic microbes (e.g., archaea) that produce methane, a much more powerful greenhouse gas than atmospheric $CO_2$. Drier conditions occurring in some locales may reduce respiratory $CO_2$ release, but favor more wildfires that disastrously destroy the metastable moss-rich environment that has persisted for millennia. In some cases, a reasonable recovery may occur within decadal time intervals; but in other cases, the scares last far longer, with much less favorable environments for plant growth or the success of indigenous and migratory animals, including birds.

The extent of these changes for human communities and human activities such as agriculture, game for hunting-dependent communities, and recreational resources that are important to the economy of some arctic regions, are not fully understood (e.g., Beaumier et al., 2015; Nicol and Heininen, 2014; Parkinson and Birgitta, 2014). However, there are sufficient early warning signs that the environmental changes have consequences for humanity that are more than of scientific interest, and locally may be more significant than broad geographic surveys can fully capture at the present time. In some locations, the thawing permafrost leads to major changes in surface topography, including thermokarst; and in some coastal regions, massive land subsidence threatens entire communities that are gradually sliding into the sea.

The changing climate of the Arctic is more than of local significance. Broad changes in surface conditions including changing snow cover patterns and effects on albedo, warming that alters the thermal balance of the tundra atmosphere, and increasingly less coastal sea ice, have ramifications for climate change globally. The Arctic is intricately linked with changing climate regimes worldwide. The manifold consequences of climate change in the Arctic ("Ground zero") for our habitable global environment are only beginning to emerge; and much more substantial research is needed to fully understand the consequences of these changes for the future of humanity on planet Earth.

## ACKNOWLEDGMENTS

My thanks to Dr. Natalie Boelman for reading the manuscript and making helpful suggestions. Some of the research by the author reviewed here was supported partially by funds from the National Science Foundation (NSF 0732664 and NSF 1043271). This is Lamont-Doherty Earth Observatory Contribution Number 7945.

## References


Adl, S. M. (2003). The Ecology of Soil Decomposition. CAB International, Wallingford, UK. p. 270–294.

Anderson, O. R. (2008). The role of amoeboid protists and the microbial community in moss-rich terrestrial ecosystems: biogeochemical implications for the carbon budget and carbon cycle, especially at higher latitudes. *Journal of Eukaryotic Microbiology*, 55, 145–150.

Anderson, O. R. (2010). The reciprocal relationships between high latitude climate changes and the ecology of terrestrial microbiota: Emerging theories, models, and empirical evidence, especially related to global warming. In: Tundras: Vegetation, Wildlife and Climate Trends, (Eds. B. Gutierrez, C. Pena). Nova Science Publ., N. Y., p. 47–79.

Anderson, O. R. (2012). The Fate of Organic Sources of Carbon in Moss-rich Tundra Soil Microbial Communities: A Laboratory Experimental Study. *Journal of Eukaryotic Microbiology*, 59, 564–570.

Anderson, O. R. (2013). Soil Respiration, Global Warming and the Role of Microbial Communities. O. R. Anderson. In: Molecular Microbial Ecology of The Rhizosphere, (Ed. F. J. de Bruijn). Wiley-Blackwell, Hoboken, N. J., p. 1055-1061.

Anderson, O. R. (2014a). The role of soil microbial communities in soil carbon processes and the biogeochemical carbon cycle. In: Soil Carbon: Types, Management Practices and Environmental Benefits (Ed. A. Margit). Nova Publishers, N. Y.

Anderson, O. R. (2014b). Survey of Bacterial and Heterotrophic Nanoflagellate Densities and C-biomass Estimates Along an Alaskan Tundra Transect with a Regression Analysis of Total Soil Respiration Predicted by Bacterial Densities. O. R. Anderson. *Journal of Eukaryotic Microbiology*, 61, 11-16.

Anderson, O. R. and Griffin, K. (2001). Abundances of protozoa in soil of laboratory-grown wheat plants cultivated under low and high atmospheric $CO_2$ concentrations. *Protistology,* 2, 76-84, 2001.

Anderson, O. R. and McGuire, K. (2013). C-biomass of bacteria, fungi, and protozoan communities in arctic tundra soil, including some trophic relationships. *Acta Protozoologica*, 52, 217-227.

Arndal, M. F., Illeris, A. L., Michelsen, A., Albert, K., Tamstorf, M., and Hansen, B. U. (2009). Seasonal variation in gross ecosystem production, plant biomass, and carbon and nitrogen pools in five high arctic vegetation types. *Arctic, Antarctic, and Alpine Research*, 41, 164-173.

Bardgett, R. D. and Wardle, D. A. (2010). Aboveground-belowground Linkages: Biotic Interactions, Ecosystems processes and global change. Oxford University Press, New York.

Barrett, K., Rocha, A. V., Van de Weg, M. J. and Gaius Shaver. (2012). Vegetation shifts observed in arctic tundra 17 years after fire. *Remote Sensing Letters*, 3, 729-736.

Beaumier, M. C., Ford, J. D. and Tagalik, S. (2015). The food security of Inuit women in Arviat, Nunaviut: the role of socio-economic factors and climate change. *Polar Record*, 51, 550-559.

Beck, P. S. A., Horning, N., Goetz, S. J., Loranty, M. M. and Tape, K. D. (2011). Shrub cover on the north slope of Alaska: a circa 2000 baseline map. *Arctic, Antarctic, and Alpine Research*, 43, 355-363.

Boelman, N. T., Gough, L., McLaren, J. R. and Greaves, H. (2011). Does NDVI reflect variation in the structural attributes associated with increasing shrub dominance in arctic tundra? *Environmental Research Letters,* 6, 035501.

Boelman, N. T., Stieglitz, M., Griffin, K. L. and Shaver, G. R. (2005). Inter-annual variability of NDVI in response to long-term warming and fertilization in wet sedge and tussock tundra. *Oecologia,* 143, 588–597.

Boelman, N. T., Stieglitz, M., Rueth, H. M., Sommerkorn, M., Griffin, K. L., Shaver, G. R. and Gamon, J. A. (2003). Response of NDVI, biomass, and ecosystem gas exchange to long-term warming and fertilization in wet sedge tundra. *Oecologia,* 135, 414–421.

Boelman, N. T., Gough, L., Wingfield, J., Goetz, S., Asmus, A., Chmura, H. E., Krause, J. S., Perez, J. H., Sweet, S. K. and Guay, K. C. (2014). Greater shrub dominance alters breeding habitat and food resources for migratory songbirds in Alaskan arctic tundra. *Global Change Biology*, 21, 1508-1520.

Borner, A. P., Kielland, K. and Walker, M. D. (2008). Effects of simulated climate change on plant phenology and nitrogen mineralization in Alaskan arctic tundra. *Arctic, Antarctic, and Alpine Research*, 40, 27-38.

Bret-Harte, M. S., Mack, M. C., Goldsmith, G. R., Sloan, D. B., DeMarco, J., Shaver, G. R., Ray, P. M., Biesinger, Z. and Chapin, III, F. S. (2008). Plant functional types do not predict biomass responses to removal and fertilization in Alaskan tussock tundra. *The Journal of Ecology*, 96, 713-726.

Bret-Harte, M. S., Mack, M. C., Shaver, G. R., Huebner, D. C., Johnston, M., Mojica, C. A., Pizano, C. and Reiskind, J. A. (2015). The response of Arctic vegetation and soils following an unusually severe tundra fire. *Philosophical Transactions of the Royal Society*: B, 368, 1-15.

Bunn, A. G. and Goetz, S. J. (2006). Trends in satellite-observed circumpolar photosynthetic activity from 1982 to 2003: The influence of seasonality, cover type, and vegetation density. *Earth Interactions*, 10, 1-19.

Callaghan, T. V. and Jonasson, S. (1995). Arctic terrestrial ecosystems and environmental-change. *Philosophical Transactions of the Royal Society A – Mathematical Physical and Engineering Sciences*, 352, 259-276.

Campbell, B. J., Polson, S. W., Hanson, T. E., Mack, M. C. and Schuur, E. A. G. (2010). The effect of nutrient deposition on bacterial communities in Arctic tundra soil. *Environmental Microbiology*, 12, 1842-1854.

Chapin III, F. S. and Shaver G. R. (1996). Physiological and growth responses of Arctic plants to a field experiment simulating climatic change. *Ecology,* 77, 822-840.

Chapin III, F. S., Shaver, G. R., Giblin, A. E., Nadelhoffer, K. J. and Laundre, J. A. (1995). Responses of Arctic tundra to experimental and observed changes in climate. *Ecology, 76*, 694-711.

Chapin, F.S., III; Sturm, M.; Serreze, M.; McFadden, J.; Key, J.; Lloyd, A.; McGuire, A.; Rupp, T.; Lynch, A.; Schimel, J. (2005). Role of land-surface changes in Arctic summer warming. *Science,* 310, 657–660.

Clarholm, M. (1981). Protozoan grazing of bacteria in soil-impact and importance. *Microbial Ecology,* 7, 343-350.

Clarholm, M. (1989). Effects of plant-bacterial-amoebal interactions on plant uptake of nitrogen under field conditions. *Biology and Fertility of Soils*, 8, 373-378.

Darbyshire, J. F. (Ed) (1994). Soil Protozoa. CAB International, Wallingford, UK.

Deslippe, J. R., Hartmann, M., Mohn, W. W. and Simard, S. W. (2011). Long-term experimental manipulation of climate alters the ectomycorrhizal community of *Betula nana* in Arctic tundra. *Global Change Biology*, 17, 1625-1636.

Deng, J., Gu, Y., Zhang, J., Xue, K., Qin, Y., Yuan, M., Yin, H., He, Z., Wu, L., Schuur, L. W. E. Tiedje, J. M. and Zhou, J. (2015). Shifts of tundra bacterial and archael communities along a permafrost thaw gradient in Alaska. *Molecular Ecology*, 24, 222-234.

Eastman, J. R., Sangermano, F., Machado, E. A., Rogan, J. and Anyamba, A. (2013). Global trends in seasonality of normalized difference vegetation index (NDVI), 1982-2011. *Remote Sensing*, 5, 4799-4818.

Elmendorf, S. C., Henry, G. H. R., Hollister, R. D., et al., (2012). Plot-scale evidence of tundra vegetation change and links to recent summer warming. *Nature Climate Change*, 2, 453-457.

Epstein, H. E., Beringer, J., Gould, W. A., Lloyd, A. H., Thompson, C. D., Chapin III, F. S., Michaelson, G. J., Ping, C. L., Rupp, T. S. and Walker, D. A. (2004). The nature of spatial transitions in the Arctic. *Journal of Biogeography,* 31, 1917-1933.

Euskirchen, E., McGuire, A. and Chapin, F. S., III. (2007). Energy feedbacks of northern high-latitude ecosystems to the climate system due to reduced snow cover during 20th century warming. *Global Change Biology*, 13, 2425–2438.

Fossøy. G., Stokke, B. G., Kasi, T. K, Dyrset, K., Espmark, Y., Hoset, K. S., Wdege, M. I. and Moksnes, A. (2014). Reproductive success is strongly related to local and regional climate in the Arctic snow bunting (*Plectrophenax nivalis*). *Polar Biology*, 38, 393-400.

Frost, G. V., Epstein, H. E., Walker, D. A., Matyshak, G. and Ermokhina, K. (2013). Patterened-ground facilitates shrub expansion in low Arctic tundra. *Environmental Research Letters*, 8, 015035.

Gamon, J. A., Huemmrich, K. F., Stone, R. S. and Tweedie, C. E. (2013). Spatial and temporal variation in primary production (NDVI) of coastal Alaskan tundra: Decreased vegetation growth following earlier snowmelt. *Remote Sensing of Environment*, 129, 144-153.

Gemi, J., Morgado, L. N., Semenova, T. A., Walker, J. M., Walker, M. D. and Smets, E. (2015). Long-term warming alters richness and composition of taxonomic and functional groups of arctic fungi. *FEMS Microbiology Ecology,* 91, DOI: http://dx.doi.org/10.1093/femsec/fiv095 fiv095.

Gough, L. and Hobbie, S. E. (2003). Responses of moist non-acidic arctic tundra to altered environment: productivity, biomass, and species richness. *Oikos,* 103, 204-216.

Gough, L., Moore, J. C., Moore, J. C., Shaver, G. R., Simpson, R. T. and Johnson, D. R. (2012). Above- and belowground responses of arctic tundra ecosystems to altered soil nutrients and mammalian herbivory. *Ecology,* 93, 1683-1694.

Grogan, P. and Chapin F. S. III (2000). Initial effects of experimental warming on above- and belowground components of net ecosystem $CO_2$ exchange in arctic tundra. *Oecologia,* 125, 512-520.

Greaves, H. E., Vierling, L. A., Eitel, J. U. H., Boelman, N. T., Magney, T. S., Prager, C. M. and Griffin, K. L. (2015). Estimating aboveground biomass and leaf area of low-stature Arctic shrubs with terrestrial LiDAR. *Remote Sensing of the Environment*, 164, 26-35.

Gustine, D. D., Brinkman, T.J., Lindgren, M. A., Schmidt, J. I., Rupp, T. S. and Adams, L. G. (2014). Climate-Driven effects of fire on winter habitat for caribou in the Alaskan-Yukon arctic, *PLOS one,* 9, e112584.

Henry, G. H. R. and Molau, U. (1997). Tundra plants and climate change: the International Tundra Experiment (ITEX). *Global Change Biology*, 3, 1-9.

Heskel, M. A, Bitterman, D., Atkin, O. K., Turnbull, M. H. and Griffin, K. (2013). Seasonality of foliar respiration in tow dominant plant species from the Arctic tundra: response to long-term warming and short-term temperature variability. *Functional Plant Biology*, 41, 287-300.

Heskel, M. A., Greaves, H. E., Turnbull, M. H., O'Sullivan, O. S., Shaver, G. R., Griffin, K. L. and Atkin, O. K. (2014). Thermal acclimation of shoot respiration in an Arctic woody plant species subjected to 22 years of warming and altered nutrient supply. *Global Change Biology,* 20, 2618-2630.

Hobbie, S. E. and Chapin, F. S. (1998). An experimental test of limits to tree establishment in Arctic tundra. *Journal of Ecology*, 86, 449-461.

Hofgaard, A., Løkken, J. O., Dalen, L. and Hytteborn, H. (2010). Comparing warming and grazing effects on birch growth in an alpine environment – a 10-year experiment, *Plant Ecology and Diversity*, 3, 19-27.

Hollister, R. D., Webber, P. J. and Tweedie, C. E. (2005). The response of Alaskan arctic tundra to experimental warming: Differences between short-and long-term responses. *Global Change Biology*, 11, 525–536.

Hufnagel, L. and Garamvölgyi, A. (2014). Impacts of climate change on vegetation distribution No. 2 – climate change induced vegetation shifts in the New World. *Applied Ecology and Environmental Research*, 12, 355-422.

Jagerbrand, A. K., Gaku, K., Alatalo, J. M. and Ulf, M. (2012). Effects of neighboring vascular plants on the abundance of bryophytes in different vegetation types. *Polar Science*, 6, 200-206.

Jandt, R., Joly, K., Meyers, R. and Racine, C. (2008). Slow recovery of lichen on burned caribou winter range in Alaska tundra: potential influences of climate warming and other disturbance factors. *Arctic, Antarctic, and Alpine Research*, 40, 89–95.

Jia, G. J., Epstein, H. E. and Walker, D. A. (2003). Greening of arctic Alaska, 1981-2001. *Geophysical Research Letters*, 30, 2067.

Joly, K., Duffy, P. A. and Rupp, T. S. (2012). Simulating the effects of climate change on fire regimes in Arctic biomes: implications for caribou and moose habitat. *Ecosphere,* 3, 36.

Jones, B. M., Kolden, C. A., Jandt, R., Abatzoglou, J. T., Urban, F. and Arp, C. D. (2009). Fire Behavior, Weather, and Burn Severity of the 2007 Anaktuvuk River Tundra Fire, North Slope, Alaska. *Arctic, Antarctic, and Alpine Research,* 41, 309-316.

Kaischke, E., Williams, D. and Barry, D. (2002). Analysis of the patterns of large fires in the boreal forest regions of Alaska. *International Journal of Wildland Fire*, 11, 131-144.

Klady, R. A., Henry, G. H. R. and Lemay, V. (2011). Changes in high arctic tundra plant reproduction in response to long-term experimental warming. *Global Change Biology*, 17, 1611-1624.

Kolden, C. A. and Rogan, J. (2013). Mapping wildfire burn severity in the arctic tundra from downsampled MODIS data. *Arctic, Antarctic, and Alpine Research*, 45, 64-76.

Koller, R., Scheu, S., Bonkowski, M. and Robin, C. (2013). Protozoa stimulate N uptake and growth of arbuscular mycorrhizal plants. *Soil Biology and Biochemistry*, 65, 204-210.

Ku, N., Popescu, S. C., Ansley, R. J., Perotto-Baldivieso, H. L. and Filippi, A. M. (2012). Assessment of available rangeland woody plant biomass with a terrestrial lidar system. *Photogrammetric Engineering and Remote Sensing,* 78, 349–361.

Lang, S. I., Cornelissen, J. H. C., Shaver, G. R., Ahrens, M., Callaghan, T. V., Molau, U., Ter Braak, C. J. F., Hölzer, A. and Aerts, R. (2012). Arctic warming on two continents has consistent negative effects on lichen diversity and mixed effects on bryophyte diversity. *Global Change Biology,* 18, 1096-1107.

Liljedahl, A., Hinzman, L., R., Busey, R. and Yoshikawa, K. (2007). Physical short-term changes after a tussock tundra fire, Seward Peninsula, Alaska. *Journal of Geophysical Research: Earth Surface*, 112, *F02S07,* DOI: 10.1029/2006JF000554.

Lin, D. H., Johnson, D. R., Andresen, C. and Weedie, C. E. (2012). High spatial resolution decade-time scale land cover change at multiple locations in the Beringian Arctic (1948-2000s). *Environmental Research Letters*, 7, 025502.

Lin, Y., Jaakkola, A., Hyyppä, J. and Kaartinen, H. (2010). From TLS to VLS: biomass estimation at individual tree level. *Remote Sensing*, 2, 1864–1879.

Lloyd, A. H., Fastie, C. L. and Eisen, H. (2007). Fire and substrate interact to control the northern range limit of black spruce (*Picea mariana*) in Alaska. *Canadian Journal of Forest Research*, 37, 2480-2493.

Loboda, T., French, N. H. F., Hight-Harf, C. and Miller, M. E. (2013). Spectral response of tundra vegetation to wildland fire. *Remote Sensing of Environment,* 134, 194-209.

Mack, M. C., Bret-Harte, M. S., Hollingsworth, T. N, Jandt, R. R., Schuur, E. A. G., Shaver, G. R. and Verbyla D. L. (2011). Carbon loss from an unprecedented Arctic tundra wildfire. *Nature,* 475, 489-492.

Mack, M. C., Schuur, E. A. G., Bret-Harte, M. S., Shaver, G. R. and Chapin, F. S. (2004). Ecosystem carbon storage in arctic tundra reduced by long-term nutrient fertilization. *Nature*, 431, 440–443.

Makoto, K. and Klaminder, J. (2012). The influence of non-sorted circles on species diversity of vascular plants, bryophytes and lichens in Sub-Arctic Tundra. *Polar Biology*, 11, 1659-1667.

Marchand, F. L., Mertens, S., Kockelbergh, F., Beyens, L. and Nijs, I. (2005). Performance of High Arctic tundra plants improved during but deteriorated after exposure to a simulated extreme temperature event. *Global Change Biology*, 11, 2078-2089.

Marcot, B. G., Jorgenson, M. T., Lawler, J. P., Handel, C. M. and DeGange, A. R. (2015). Projected changes in wildlife habitats in Arctic natural areas of northwest Alaska. *Climate Change*, 130, 145-154.

McLaughlin, B. C., Xu, C.-Y., Rastetter, E. B. and Griffin, K. L. (2014). Predicting ecosystem carbon balance in a warming Arctic: the importance of long-term thermal acclimation potential and inhibitory effects of light on respiration. *Global Change Biology*, 20, 1901-1912.

Morgado, L. N., Semenova, T. A., Welker, J. M., Walker, M. D., Smets, E. and Geml, J. (2015). Summer temperature increase has distinct effects on the ectomycorrhizal fungal communities of moist tussock and dry tundra in Arctic Alaska. *Global Change Biology*, 21, 959-972.

Moyes, A. B., Castanaba, C., Germino, M. J. and Kueppers, L. M. (2013). Warming and the dependence of limber pine (*Pinus flexilis*) establishment on summer soil moisture within and above its current elevation range. *Oecologia,* 171, 271-282.

Myers-Smith, I. H., Forbes, B. C., Wilmking, M., Hallinger, M. Lantz, T. Blok, D. Tape, K. D., et al., (2011). Shrub expansion in tundra ecosystems: dynamics, impacts and research priorities. *Environmental Research Letters*, 6, 045509.

Naito, A. T. and Cairns, D. M. (2011). Patterns and processes of global shrub expansion. *Progress in Physical Geography*, 35, 423-442.

Narita, K., Harada, K., Saito, K. Sawada, Y., Fukuda, M. and Tsuyuzaki, S. (2015). Vegetation and permafrost thaw depth 10 years after a tundra fire in 2002, Seward Peninsula, Alaska. *Arctic Antarctic and Alpine Research*, 47, 547-559.

Nicol, H. N. and Heininen, L. (2014). Human security, the Arctic Council and climate change: competition or co-existence? *Polar Record*, 50, 80-85.

Nicolson, C., Berman, M., West, C. T., Kofinas, G. P., Griffith, B., Russell, D. and Dugan, D. (2013). Seasonal climate variation and caribou availability: modeling sequential movement using satellite-relocation data. *Ecology and Society*, 18, 1. http://dx.doi.org/10.5751/ES-05376-180201.

Oberbauer, S. F., Elmendorf, S. C., Troxler, T. G., Hollister, R. D., Rocha, A. V., Bret-Harte, M. S., Dawes, M. A., Fosaa, A. M., Henry, G. H. R., Høye, T. T., Jarrad, F. C., Jónsdóttir, I. S., Klanderud, K., Klein, J. A., Molau, U., Rixen, C., Schmidt, N. M., Shaver, G. R., Slider, R. T., Totland, Ø., Wahren, C.-H. and Welker, J. M. (2013). Phenological response of tundra plants to background climate variation tested using the International Tundra Experiment. The Royal Society, *Philosophical Transactions B*, 368, 20120481.

Old, K. M., Darbyshire J. F. (1978). Soil fungi as food for giant amoebae. *Soil Biology and Biochemistry*, 10, 93–100.

Old, K. M., Darbyshire J. F. (1980) Arachnula impatiens Cienk., a mycophagous giant amoeba from soil. *Protistologica,* 16, 277–287.

Osterkamp, T. E., Jorgenson, M. T., Schuur, E. A. G., Shur, Y. L., Kanevskiy, M. Z., Vogel, J. G. and Tumskoy, V. E. (2009). Physical and ecological changes associated with warming permafrost and thermokarst in interior Alaska. *Permafrost and Periglacial Process*, 20, 235-256.

Olsoy, P. J., Glenn, N. F., Clark, P. E. and Derryberry, D. R. (2014). Aboveground total and green biomass of dryland shrub derived from terrestrial laser scanning. ISPRS *Journal of Photogrammetry and Remote Sensing,* 88, 166–173.

Parkinson, A. J. and Birgitta, E. (2014). In: Butler, C. D. (ed.), Climate Change and Global Health. CAB International, Wallingford, U.K., p. 206-217.

Pattison, R. R., Jorgenson, J. C., Raynolds, M. K. and Welker, J. M. (2015). Trends in NDVI and tundra community composition in the Arctic of NE Alaska Between 1984 and 2009. *Ecosystems,* 18, 707-719.

Park, H. J., Chae, N., Sul, W. J. Lee, B. Y., Lee, Y. K. and Kim, D. (2015). Temporal changes in soil bacterial diversity and humic substances degradation in subarctic tundra soil. *Microbial Ecology*, 69, 668-675.

Post E., Mads C. Forchhammer, M. Bret-Harte, S., Callaghan, T. V., Christensen, T. R., Elberling, B., Fox, A. D., Gilg, O., Hik, D. S., Høye, T. T., Ims, R. A., Jeppesen, E., Klein, D. R., Madsen, J., McGuire, A. D., Rysgaard, S., Schindler, D. E., Stirling, I., Tamstorf, M. P., Tyler, N., van der Wal, R., Welker, J., Wookey, P. A., Schmidt, M. and Aastrup, P. (2009). Ecological dynamics across the Arctic associated with recent climate change. *Science,* 325, 1355-1358.

Potter, C. (2014). Regional analysis of MODIS satellite greenness trends for ecosystems of interior Alaska. GIscience and Remote Sensing, 51, 390-402.

Racine, C., Jandt, R., Meyers, C. and Dennis, J. (2004). Tundra fire and vegetation change along a hillslope on the Seward Peninsula, Alaska, U.S.A. *Arctic, Antarctic, and Alpine Research*, 36, 1-10.

Raynolds, M. K., Walker, D. A., Verbyla, D. and Munger C. A. (2013). Patterns of change within a tundra landscape: 22-year Landsat NDVI trends in an area of the Northern Foothills of the Brooks Range, Alaska. *Arctic, Antarctic, and Alpine Research*, 45, 249-260.

Rocha, A. V. and Shaver, G. (2011). Post-fire energy exchange in arctic tundra: the importance and climatic implications of burn severity. *Global Change Biology*, 17, 2831-2841.

Rønn, R., Ekelund, F. and Christensen, S. (2003). Effects of elevated atmospheric $CO_2$ on protozoan abundances in soil planted with wheat and on decomposition of wheat roots. *Plant and Soil,* 251, 13-21.

Rønn, R., Gavito, M., Larsen, J., Jakobsen, I., Frederiksen, H. and Christensen, S. (2002). Response of free-living soil protozoa and microorganisms to elevated $CO_2$ and presence of mycorrhiza. *Soil Biology and Biochemistry*, 34, 923-932.

Rumpf, S. B., Semenchuk, P. R., Dullinger, S. and Cooper E. J. (2014). Idiosyncratic Responses of high arctic plants to changing snow regimes. *PLoS ONE* 9(2): e86281.

Schuur, E. A. G., Crummer, K. G., Vogel, J. G. and Mack, M. C. (2007). Plant species composition and productivity following permafrost thaw and thermokarst in Alaskan tundra. *Ecosystems,* 10, 280-292.

Semenova, T. A., Morgado, L. N., Welker, J. M., Walker, M. D., Smets, E. and Geml, J. (2015). Long-term warming alters community composition of *ascomycetes in Alaskan moist and dry arctic tundra. Molecular* Ecology, 24, 424-437.

Shaver, G. R., Bret-Harte, M. S., Jones, M. H., Johnstone, J., Gough, L., Laundre, J. and Chapin III, F. S. (2001). Species composition interacts with fertilizer to control long-term change in tundra productivity. *Ecology,* 82, 3163-3181.

Stieglitz, M., McKane, M. and Klausmeier, C. A. (2006). A simple model for analyzing climatic effects on terrestrial carbon and nitrogen dynamics: An arctic case study. *Global Biogeochemical Cycles*, 20, GB3016.

Sturm, M., Racine, C. and Tape, K. (2001). Climate change: Increasing shrub abundance in the Arctic. *Nature,* 411, 546-547.

Sweet, S. K., Griffin, K. L., Steltzer, H., Cough, L. and Boelman, N. (2015). Greater deciduous shrub abundance extends tundra peak season and increases modeled net $CO_2$ uptake. *Global Change Biology*, 21, 2394-2409.

Tape, K. D., Hallinger, M., Welker, J. M. and Ruess, R. W. (2012). Landscape heterogeneity of shrub expansion in arctic Alaska. *Ecosystems,* 15, 711–24.

Tape, K. D., Sturm, M. and Racine C. (2006). The evidence for shrub expansion in Northern Alaska and the Pan-Arctic. *Global Change Biology,* 12, 686-702.

Timling, I. and Taylor, D. L. (2012). Peeking through a frosty window: molecular insights into the ecology of Arctic soil fungi. *Fungal Ecology*, 5, 419-429.

Treonis, A. M. and Lussenhop, J. F. (1997). Rapid response of soil protozoa to elevated $CO_2$. *Biology and Fertility of Soils,* 25, 60-62.

Van Oudenhove, L., Gauthier, G. and Lebreton, J.-D. (2014). Year-round effects of climate on demographic parameters of an arctic-nesting goose species. *Journal of Animal Ecology*, 83, 1322-1333.

Van Wijk, M. T., Clemmensen, K. E., Shaver, G. R., Williams, M., Callaghan, T.V., Chapin III, F. S., Cornelissen, J. H.C., Gough, L., Hobbie, S. E., Jonasson, S., Lee, J. A., Michelsen, A., Press, M. C., Richardson, S. J. and Rueth, H. (2003). Long-term ecosystem level experiments at Toolik Lake, Alaska and at Abisko, Northern Sweden: generalizations and differences in ecosystem and plant type responses to global change. *Global Change Biology,* 10, 105-123.

Verbyla, D. (2008). The greening and browning of Alaska based on 1982-2003 satellite data. *Global Ecology and Biogeography*, 17, 547-555.

Vierling, L. A., Xu, Y., Eitel, J. U. H. and Oldow, J. S. (2012). Shrub characterization using terrestrial laser scanning and implications for airborne LiDAR assessment. *Canadian Journal of Remote Sensing*, 38, 709–722.

Vors, L. S. and Boyce, M. S. (2009). Global declines in caribou and reindeer. *Global Change Biology*, 15, 2626-2633.

Wahren, C.–H. A., Walker, M. D. and Bret-Harte, M. S. (2005). Vegetation responses in Alaskan arctic tundra after 8 years of a summer warming and winter snow manipulation experiment. *Global Change Biology*, 11, 537-552.

Wallenstein, M. D., McMahon, S. and Schimel, J. (2007). *FEMS Microbiology and Ecology,* 59, 428-435.

Walker, D. A., Epstein, H. E., Jia, G. J., Balser, A., Copass, C., Edwards, E. J., Gould, W. A., Hollingsworth, J., Knudson, J., Maier, H. A., Moody, A. and Raynolds, M. K. (2003). Phytomass, LAI, and NDVI in northern Alaska: Relationships to summer warmth, soil pH, plant functional types, and extrapolation to the circumpolar Arctic. *Journal of Geophysical Research*, 108, 8169.

Walter, G. R. (2004). Plants in a warmer world. *Perspectives in Plant Ecology Evolution and Systematics*, 6, 169-185.

Wild, T. C., Kendall, S. J., Guldager, N. and Powell, A. N. (2015). Breeding habitat associations and predicted distribution of an obligate tundra-breeding bird, Smith's Longspur. *The Condor*, 117, 3-17.

Wheeler, H. C., Chipperfield, J. D., Roland, C. and Svenning, J.-C. (2015). How will the greening of the Arctic affect an important prey species and disturbance agent? Vegetation effects on arctic ground squirrels. *Global Change Ecology*, 178, 915-929.

Xu, L., Myneni, R. B., Chapin III, F. S., Callaghan, T. V., Pinzon, J. E., Tucker, C. J., Zhu, Z., Bi, J., Ciais, P., Tømmervik, H., Euskirchen, E. S., Forbes, B. C., Piao, S. L., Anderson, B. T., Ganguly, S., Nemani, R. R., Goetz, S. J., Beck, P. S. A., Bunn, A. G., Cao, C. and Stroeve, J. C. (2013). Temperature and vegetation seasonality diminishment over northern lands. *Nature Climate Change*, 3, 581–586.

Yarie, J. (1981). Forest fire cycles and life tables: a case study from interior Alaska. *Canadian Journal of Forest Research*, 13, 721-728.

Zeng, H., Jia, G. and Epstein, H. (2011). Recent changes in phenology over the northern high latitudes detected from multi-satellite data. *Environmental Research Letters*, 6, 045508.

In: Terrestrial Biomes
Editor: Marlon Nguyen
ISBN: 978-1-63484-625-7


*Chapter 3*

# SPONTANEOUS STAND REGENERATION AND HERB LAYER RESTORATION IN POST-FIRE WOODS 16 YEARS AFTER A FOREST FIRE (RUDZINIEC FORESTS, SOUTHERN POLAND CASE)

***Anna Orczewska, Katarzyna Żołna and Małgorzata Żaczek***
University of Silesia, Faculty of Biology and Environmental Protection,
Department of Ecology, Katowice, Poland

## ABSTRACT

A survey of the development of the stand structure and the recovery of the herb layer in 16-year-old forests that appeared spontaneously after a forest fire was undertaken in southern Poland where the forests were burnt in 1992. Stand structure and its natural regeneration were surveyed in thinned and unthinned forests. The heights and diameters of trees were measured within each of the ten 40 x 20 m plots. In addition, ten transects (10 m long by 2 m wide, consisting of 5 quadrats, $4m^2$ each) were set up in each of them. In each 2 x 2 m plot, the number of seedlings of tree and shrub species up to 0.5 m high, 0.5-2.0 m and over 2 m high that had a diameter < 5 cm were counted. The tree canopy cover was measured in each 2x2 m quadrat. Then, the herb layer composition in thinned and unthinned forests was studied. Thus, in randomly located 10x10 m plots (61 in the managed and 61 in unthinned forests), the percentage cover of herbaceous vascular plant species was estimated. Then, a numerical classification of the plots was conducted. The effect of thinning on stand composition and on the number of trees was very significant; the stand in thinned forest was exclusively composed of silver birch (density – 1,688-3,112 individuals per hectare), whereas in the unthinned forest the birch density reached 1,225-1,900 individuals per hectare and six other tree species were also present in this stand. Both forests differed in their canopy cover and the diameter of the birch trees (significantly higher in the thinned forests and lower in the unthinned forests, respectively). The response of a stand to thinning was also detectable in its regeneration process, which was more dynamic in the unthinned forest. A reaction to thinning was also noticed when the horizontal structure of the herb layer was compared. The most important difference was related to the increased level of the illumination of the forest floor within the thinned stand, which resulted in the increased abundance of different species of grasses and sedges, as compared to the herb layer in an unthinned stand. The main conclusion is that although there is a potential to provide timber of a better quality in the case of thinned forests, the recolonization of the herb layer by true forest species is slower than in successional forests that lack any management measures of their stands.

## INTRODUCTION

In many parts of the world fire is regarded as a natural disturbance factor in many types of vegetation communities that contributes to temporal changes in community composition. In such fire-prone communities, "plant population changes display episodes of rapid population growth or decline rather than gradual, directional changes" in response to the cyclic appearance of fire (Whelan 1995). The 'blind' application of the terminology that is associated with the succession theory to describe post-fire vegetation changes that take place in fire-prone ecosystems has been widely criticized since processes other than succession are much more responsible for community changes after a fire in such ecosystems. In contrast, in fire-sensitive plant communities that are affected by a single fire, a successional process of directional community development is initiated. In such a situation, the pattern of the changes in the community resembles the one that has been described in the case of vegetation development on abandoned fields in many aspects. Thus, in general, the species richness and species composition turnover during the development pattern of a post-fire community can be regarded as an example of secondary succession (Whelan 1995).

Alteration of the physical environment, which is one of the first order effects of fire, has great implications on the successive regeneration and development of vegetation cover. Among the changes in physical conditions resulting from fire, an increase in the range of soil temperatures, especially the maximum ones, is an important factor. This happens as a consequence of the openness of a site because most of the aboveground vegetation is destroyed and is also due to the removal/burn of the litter and organic layers and as a result of an altered albedo of the soil surface (Whelan 1995; Certini 2005). Removal of the aboveground biomass leads to an increased wind speed, which in turn contributes to stronger wind erosion and the successive drying of the upper layer of soil. Another consequence of fire is an alteration of water conditions, including a changed water storage capacity and pattern of runoff. A higher level of evaporation due to higher temperature and wind speed and increased water loss due to erosion, which are a result of the lack of a humus layer and vegetation cover, are observed (Whelan 1995; Brown and Smith 2000; Certini 2005). As a result, altered post-fire conditions make the process of forest recovery very difficult.

Temperate deciduous and mixed coniferous-broadleaved forests in the northern hemisphere belong to fire-sensitive plant communities, and therefore, fire is not treated as a cyclic disturbance but is regarded as a catastrophe to the forest ecosystem. For these reasons plants in such communities did not develop any evolutionary adaptations to fire. In the event of high-intensity fires, therefore, most, if not all of the established plant populations of forest species are killed (Whelan 1995). This makes the process of forest recovery, which relies mostly on seed dispersal from outside, very slow, especially when compared with fire-prone communities where many species germinate from the buried seeds once fire is over (Whelan

1986; cited by Whelan 1995). Since fire in central Europe is not a natural disturbance factor, studies on the effect of single fires on vegetation communities are very limited (Faliński 1998a; Kwiatkowska-Falińska 2008; Budzáková et al. 2013). This is especially the case with post-fire forest succession on sites that were also forested prior to a fire. Although such studies are very important, mainly due to the fact that the fire risk here is growing due to current changes in weather patterns, which is clearly visible in increasing periods of heat and drought and due to expected climate changes in the nearest future, they are not yet very common. Thus, taking the advantage of the fact that in the early 1990s, great areas of forest in southern Poland were burnt, we decided to study the state of forest community development on these sites 16 years after the fire. Therefore, we focused on the stand composition and regeneration and on the horizontal structure of the herbaceous layer of these communities.

(a) (b)

Figure 1 a,b. General views of the fire (photos from the archives of the Rudy Raciborskie Forest District).

## SHORT HISTORY OF THE SITE

The studies were conducted in Kotlina Raciborska, Upper Silesia, Rudziniec Forest District, Łącza Subdistrict in southern Poland, where one of the most extensive and intensive forests fires, which was ignited by a man, occurred in central Europe in the summer of 1992 (Figures 1a, b). It happened in late August during an extremely hot (28-31.3°C) and dry (20% relative humidity) period of weather, which was accompanied by strong winds (24m/sec). To make the situation worse, prior to fire there was a rather long period of drought that had also resulted from weather conditions. Such a weather pattern made it very difficult to conquer the fire, and therefore, the direct battle with the flames lasted six days and which was followed by over 20 days of dealing with local burns. The disaster covered the area of 9,062 hectares, which were completely burnt, including over 5,000 hectares of mature stands, predominantly pine-oak mixed fresh and humid forests (Hanak 1994a, b; Szabla 1994). Forests belonging to three administrative districts that are governed by Polish State Forests were affected; 4,480 hectares in the Rudy Raciborskie District, 2,352 hectares in Rudziniec and 2,230 hectares in the Kędzierzyn Forest District (Szabla 1994; Fronczak 2012). This high-intensity ground and stand fires led to dramatic alterations in the physio-chemical conditions of the site, alterations in the soil temperature, soil moisture levels, insolation and nutrient availability. Among others, these transformations involved a total burnout of the litter and humus layers to a depth of 10-30 cm, and consequently, an almost total loss of soil enzyme activity (Hanak 1997;

Olszowska 2002), followed by the loss of ash due to very strong wind erosion, a drop of pH level to 3-4, a decrease in the amount of carbon, nitrogen and phosphorus, a drop in the groundwater level (Cieplowski and Stolarek 1995; Zwoliński et al. 2004), a change in the microclimatic conditions, along with an increase in the incidents of extremely high and low temperatures and strong winds. The latter contributed to the frequent events of sand and ash storms (Figure 2) that took place for many months after the fire, especially in the spring of 1993 (Hanak 1994a, b; Szabla 1994; Hanak 1997; Hawrys 1998; Fronczak 2012). Planning an appropriate management plan to hasten the recovery process of such a large area that had totally burnt was a great challenge for foresters. Due to the loss of any remaining old woods that could have potentially played the role of seed sources for tree species, trees had to be planted in much of the area (Bernadzki and Brzeziecki 1999). The management of forests that are owned by the State Forests in Poland is governed by The Rules of Forest Management document and according to these rules, a manager is obliged to reforest a site where a forest was lost as a result of a natural catastrophe before the fifth year after the disturbance (Rozwałka 2003; Dobrowolska 2008). In this respect, there was an urgent need for a strategy on how to hasten the recovery of post-fire sites on such large areas that had been totally denuded by the disturbance. Then, the strategy, which involved the preparation of a site followed by a tree planting campaign of carefully selected species for that purpose, depended on the type of forest site. In poor habitats, *Pinus sylvestris* (Scots pine) was chosen most often, whereas in richer sites either spontaneous colonization by *Betula pendula* (silver birch) took place (which was already observed in the spring of 1993) or *Larix decidua* (European larch) was planted. In the Rudy Raciborskie Forest District, trees (mainly Scots pine) were planted on an area of 3,000 hectares and the remaining 1,500 hectares were stands that originated from the natural succession of the pioneer tree species, mainly silver birch. In the Kędzierzyn Forest District in 1996, 1,700 (mostly Scots pine) and 441 hectares (mostly silver birch) were planted, whereas in the Rudziniec Forest District, where forests that were predominantly on more fertile habitats were affected by fire, over 1,137 hectares of post-fire sites were reforested and another 911 hectares were covered by post-fire forests that occurred spontaneously as a result of secondary succession. The latter ones were primarily composed of silver birch or, in more humid sites, black alder *Alnus glutinosa* (Fronczak 2007). Thus, in the case of the Rudziniec Forest District, spontaneous early successional communities that were composed of silver birch reached 48% of the post-fire sites. Stands of such an origin facilitated the recovery process of the burnt areas since they did not require any planting but instead humans took advantage of the dynamic, natural processes that occurred in nature. The successive appearance of such extensive areas being colonized naturally by pioneer tree species created a good opportunity for managers to exclude at least part of these forests from active silvicultural management practices. Such a decision permitted the observation of the process of natural forest dynamics and the directions of stand regeneration. For this purpose, 32.02 hectares of forests that had stands that appeared spontaneously, which were mainly composed of two typical seeders, either silver birch or aspen *Populus tremula*, that were located in the Rudziniec Forest District were selected (Fronczak 2012). It is not a large area compared to the total area that was covered by young stands that originated from natural succession that were available there. Nevertheless, today, two decades after the forest fire, this small area provides ecologists with a great opportunity to observe how the forest regeneration has progressed so far and how the dynamics of the species growth and turnover will continue in the future. Unfortunately, since no widely planned and carefully designed research was conducted in the first years after the fire (no information about the forest regeneration process at the starting point and no grid of plots for permanent, long-term studies are available), there is a gap in knowledge in this respect. Only a few studies on the stand dynamics were undertaken in the neighboring Rudy Raciborskie Forest District, but most of these were conducted more than ten years after the fire and did not include all of the forest site types that are covered by the current post-fire forest. For these reasons no data exist that could be used for comparison with the results that we obtained.

Figure 2. An ash storm, which frequently occurred in the area of a burned forest (a photo from the archives of the Rudy Raciborskie Forest District).

## Field Methods and Data Analyses

In 2008, a survey was conducted in 16-year-old post-fire forests with a stand that had originated from spontaneous succession that belonged to the Rudziniec Forest District, Łącza Subdistrict (N50$^{0}$20'07"; E18$^{0}$26'55"). Out of the total area of the Łącza Subdistrict of about 683.16 hectares, 644.71 hectares (94%) of forests were completely burned, whereas very young forests were predominant in the stands that survied (15.67 ha). Three forest sections were selected for the studies (Figure 3). Among them, numbers 153a and b with an area of 18.98 hectares in total represented the stands that had appeared spontaneously after the fire and did not undergo any silvicultural treatments but were left unmanaged (untouched) in order to observe the natural processes of forest dynamics (Figure 4a). The third forest section that was selected for the investigations, No. 154a, which covers an area of 21.2 hectares, was covered by silver birch stands that had resulted from the natural succession but that were actively managed afterwards (Figure 4b). The silvicultural practices that were conducted within this area included early, selective cleaning (juvenile thinning), which was done manually in 1999-2000 (7-8 years after the fire), the aim of which was to eliminate *Populus tremula* seedlings and to reduce the average density of birch seedlings from 30 to 3-4 individuals per m$^2$. Both types of stands that developed on the site of the burned forests were originally dominated by fifty to eighty-year-old *Pinus sylvestris* (sections No 154a and 153a, respectively) or, in the case of section No 153b, with forty-year-old *Fagus sylvatica* stands. These stands grew on a habitat of fresh mixed broadleaved forest site and on fresh and moist broadleaved forest sites, respectively, and therefore the sites were too rich for Scots pine. Prior to fire, the soils of these sites were classified as Dystric Cambisols and Albic Luvisols (Plan Urządzania 1985).

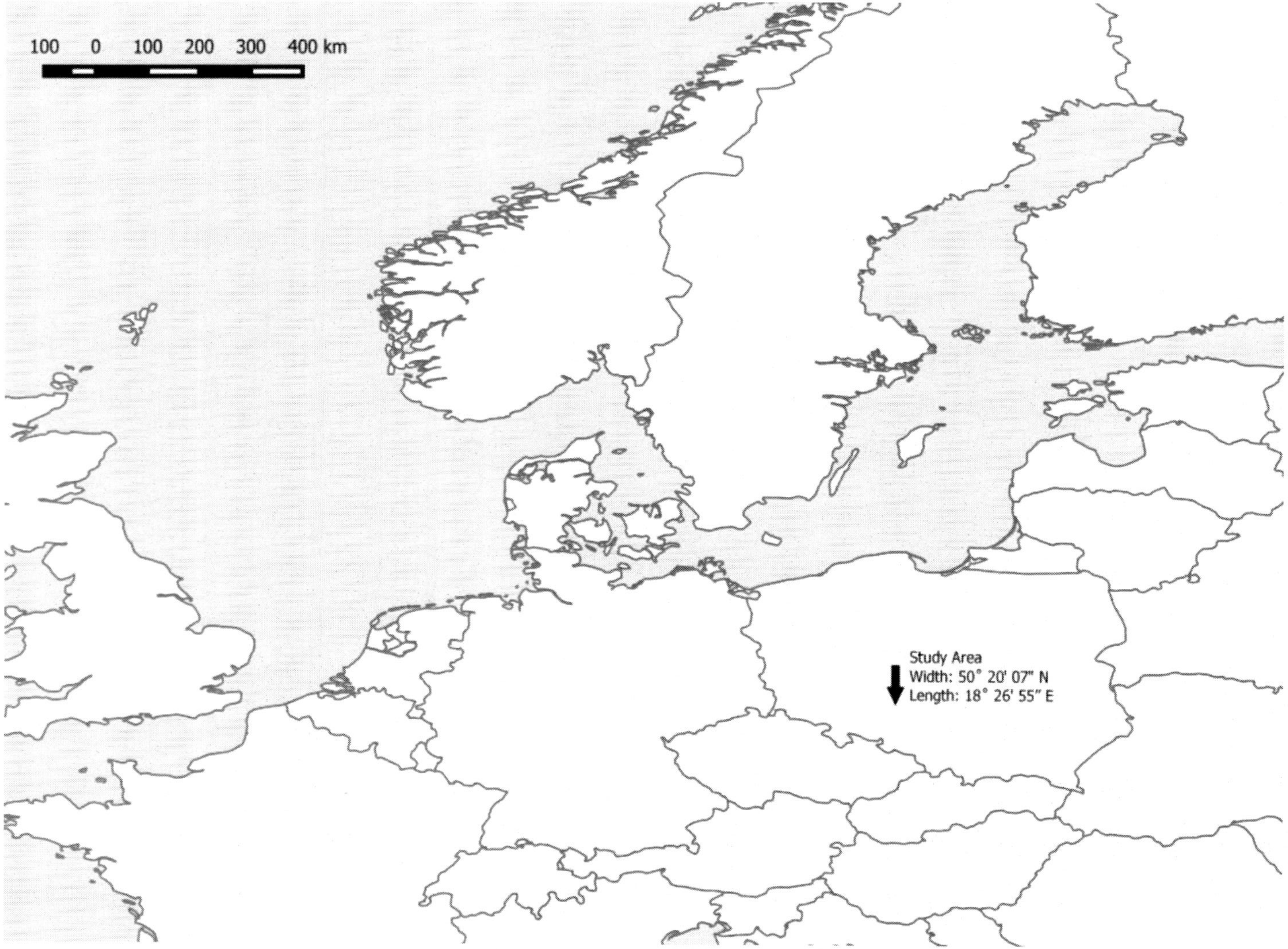
100 0 100 200 300 400 km
Study Area
Width: 50° 20' 07" N
Length: 18° 26' 55" E

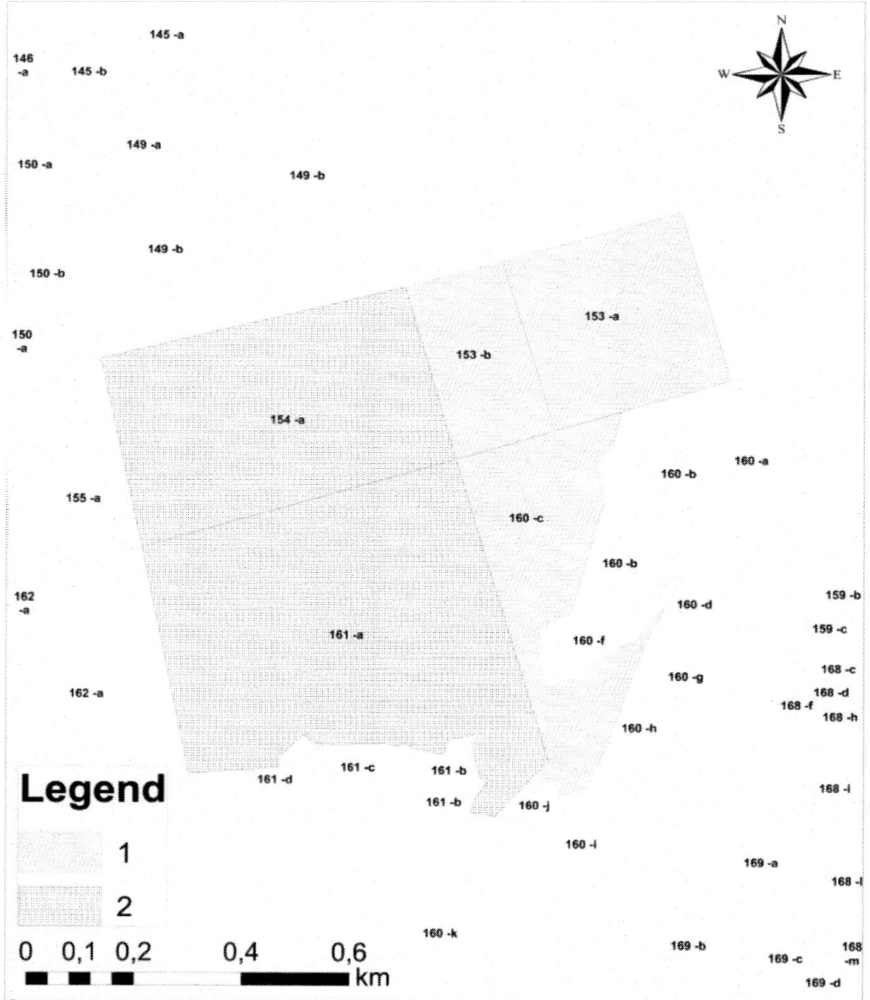


Figure 3. The location of the study site. 1 – unthinned forest, 2 – thinned forest.

Figure 4. General view of the a) unthinned and b) thinned post-fire forest (Photo by Anna Orczewska).

## Stand Structure and Its Natural Regeneration

Following the methods proposed by Jaworski and Nosek (1983), the survey of the stand structure and the natural regeneration of woody species was carried out within ten randomly located 20x40 meter sampling plots, five in the thinned (plots I_t – V_t) and five in the forests that had not been subjected to any thinning treatments (plots VI_n-t – X_n-t). The heights and diameters at breast height of all trees were measured in each of the ten plots and the presence of all shrub species was recorded. In addition, ten transects (10 m long by 2 m wide, consisting of five quadrats of $4m^2$) were set up systematically in each 20 x 40 m plot according to the scheme presented on Figure 5. This number of transects represented 25% of the area of the 20 x 40 m plot, and therefore was sufficiently representative of the whole sampling area of 800 $m^2$. In each 2 x 2 m plot (in total 50 quadrats per each 20 x 40 m plot), the number of tree and shrub species seedlings up to 0.5 m high, 0.5-2.0 m and over 2 m high that had a diameter of less than 5 cm were counted. In addition, the total herb layer cover was visually estimated in each 2 x 2 m quadrat and the mean tree canopy cover was measured using a spherical densiometer (Model A) – the mean value of four readings from the densiometer per 2 x 2m plot was determined.

The number of each of the tree species that were present in the 20 x 40 plots of the thinned and unthinned forests was counted and then the density of the individuals per hectare was calculated. The differences in the mean values of the tree canopy cover and the cover of herbaceous layer between the forest types that were studied and in all of the variables that characterized the stand structure and regeneration were tested for significance using the Mann-Whitney U test (Statistica 10.0 package).

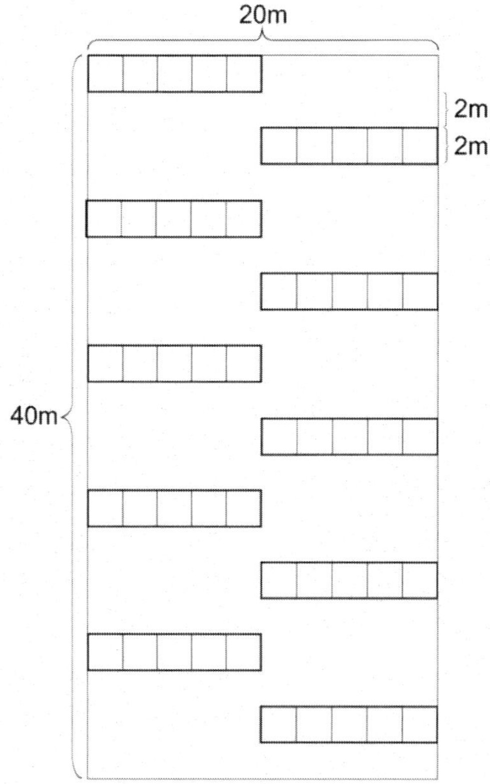


Figure 5. Scheme of the plots for the studies of the stand composition and regeneration.

## Herbaceous Layer Composition

As the next step the herb layer composition in both forest types was studied. In order to do this, lists of all of the species of vascular plants that were present in the herb layer were prepared and then their percentage of cover, according to the following scale: up to 1%, 5%, 10%, 20%... and 100%, was estimated for 61 randomly located 10x10 m plots in the thinned and 60 randomly located 10 x 10 m plots in the unthinned forests,. In addition, a visual estimation of the total cover of shrub layer was done in each plot and the tree canopy cover was measured using a spherical densiometer (mean of four readings per plot). Then, in order to examine the variation in the species composition of the herb layer and to distinguish the microcommunities that reflected the horizontal structure of that layer, a numerical classification of the plots was conducted. The Euclidean distance was applied as the measure of the plots' similarity and the minimal variance as the method of their clustering (using MVSP – MultiVariate Statistical Package, version 3.1). This analysis was done separately for the quadrats that had originated from the thinned and unthinned forests. Additionally, the Shannon diversity index per each plot was calculated with the same statistical package. The nomenclature of vascular plant species follows Mirek et al. (2002), the classification of species to forest and non-forest groups was based on Matuszkiewicz (2001) and the nomenclature of the microcommunities is based on the names of the dominant species.

For each herb layer species its classification according to the Raunkiaer life forms, Grime life strategies and seed dispersal modes is given based on the information about species traits that were obtained from the LEDA Traitbase (Kleyer et al. 2008). In the next step, significant differences between the forests in their mean number and cover of herb layer species, the mean value of the Shannon diversity index, the mean frequency and cover of forest and non-forest species and the species that represented the different traits listed above were tested using the Mann-Whitney U test. Differences in the same variables but among the microcommunities in the thinned and unthinned forests that were distinguished in the course of the classification were also tested using the Kruskal-Wallis test. In addition, lists of the herb layer species that show the affinity to each of the two forest types studied were created. The Fisher exact probability test was applied to do this. All calculations were done using the Statistica 10.0 package.

**Table 1. Tree species composition and density (number of individuals per hectare) and the mean tree and herb layer cover on the plots of the thinned (I_t – V_t) and unthinned (VI_n-t – X_n-t) post-fire forests**

| Plot symbol | Density of individuals per hectare | | | | | | | Mean (SD) tree layer cover | Mean (SD) herb layer cover |
|---|---|---|---|---|---|---|---|---|---|
| | ***Betula pendula*** | ***Populus tremula*** | ***Frangula alnus*** | ***Quercus robur*** | ***Quercus petraea*** | ***Larix decidua*** | ***Pinus sylvestris*** | | |
| I_t | 3,112 | 0 | 0 | 0 | 0 | 0 | 0 | 86.04 (8.22) | 95.92 (5.58) |
| II_t | 2,225 | 0 | 0 | 0 | 0 | 0 | 0 | 86.91 (7.69) | 91.36 (18.88) |
| III_t | 2,175 | 0 | 0 | 0 | 0 | 0 | 0 | 85.55 (5.50) | 89.70 (16.18) |
| IV_t | 1,688 | 0 | 0 | 0 | 0 | 0 | 0 | 73.18 (9.81) | 81.40 (28.39) |
| V_t | 2,100 | 0 | 0 | 0 | 0 | 0 | 0 | 81.88 (10.92) | 72.40 (31.95) |
| VI_n-t | 1,900 | 525 | 12 | 0 | 0 | 0 | 0 | 94.83 (5.61) | 75.40 (28.32) |
| VII_n-t | 1,275 | 312 | 0 | 84 | 0 | 12 | 24 | 96.42 (2.93) | 57.00 (29.81) |
| VIII_n-t | 1,225 | 175 | 0 | 72 | 60 | 12 | 12 | 95.92 (4.27) | 65.80 (30.98) |
| IX_n-t | 1,525 | 287 | 0 | 0 | 0 | 0 | 0 | 96.43 (2.60) | 68.60 (27.66) |
| X_n-t | 1,437 | 225 | 0 | 0 | 0 | 0 | 0 | 97.82 (2.35) | 57.80 (16.32) |

## RESULTS

### Stand Structure and its Natural Regeneration in Thinned and Unthinned Post-Fire Forests

The stand in forest that was subjected to thinning was exclusively composed of silver birch (Figure 4b). The number of tree individuals per hectare varied from 1,688 to 3,112 (Table 1). The height of the trees was almost uniform and ranged from 11 to 13.5 meters. The diameters of the trees were more diverse and ranged from 4.6 to 14.8 cm, with 6-8 cm wide individuals being the most frequent (median = 7.2 cm). The detailed spectrum of the tree diameters within the plots of the thinned forest is given in Figure 6.

The forest with the unthinned stand was not purely composed of silver birch but it had a diverse composition (Figure 7), which consisted of seven species. Among them, silver birch was the most abundant (1,225-1,900 individuals per hectare) and *Populus tremula* was the second most frequent (175-525 individuals per hectare), whereas the remaining five species, with the exception of *Quercus robur*, occurred rather incidentally within the plots (Table 1). The spectrum of tree diameters at breast height, especially for *Betula pendula* and *Populus tremula*, was also diverse. In the case of silver birch, the most frequent values ranged from 5 to 7 cm (median 6.5; mean 6.94 cm) and for the second species 6 to 10 cm (median 8.4; mean 8.95) (Figures 8-9). The mean and median values of the diameters of the *Quercus robur* individuals was 7.8 and 7.5 cm, 7.76 and 7.1 cm, respectively, in the case of *Q. petraea* and 13.92 and 11.9 cm for *Pinus sylvestris*. The single individuals of *Larix decidua* and *Frangula alnus* were 7.5 and 6.2 cm wide, respectively.

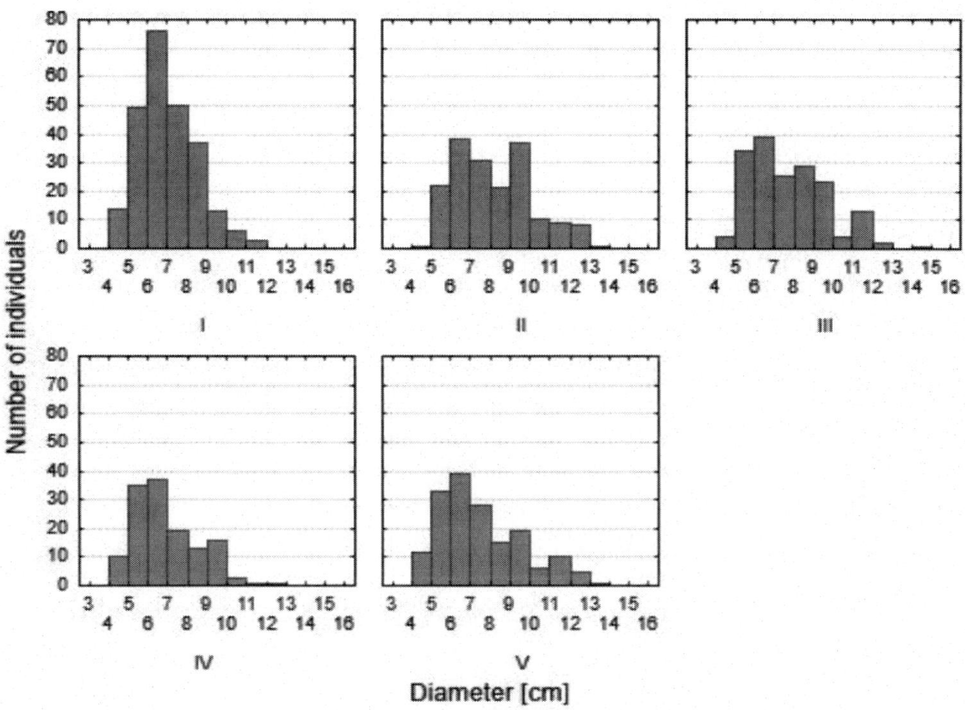


Figure 6. The diameter range values for *Betula pendula* on the plots of the stand in the thinned forest.

Figure 7. A diverse stand of the unthinned forest (Photo by Anna Orczewska).

The forests differed significantly in the mean cover of the tree and herb layers (at $p < 0.05$; according to the Mann-Whitney U test). The tree canopy cover was higher in the unthinned forest than in the one that was subjected to thinning (96.29 [± 3.85 SD]% and 82.71 [± 9.95 SD]%, respectively) (Figures 10-11), whereas in the case of the herbaceous layer, the situation was reverse (86.16 [± 23.60 SD]% cover in the thinned and 64.92 (± 27.79 SD)% in the unthinned forest). Significant differences were also recorded in the mean diameter of *Betula pendula*, which was larger in the case of the thinned forest (7.52 [± 1.88 SD] cm versus 6.9 [± 1.59 SD] cm in the unthinned forest).

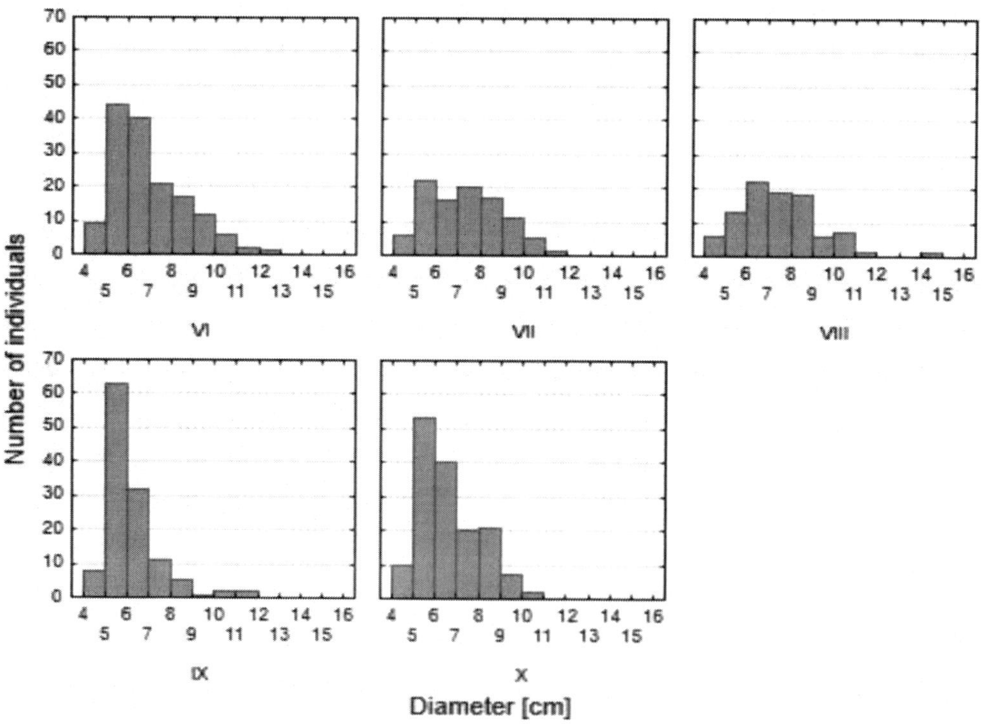


Figure 8. The diameter range values for *Betula pendula* on the plots of the stand in the unthinned forest.

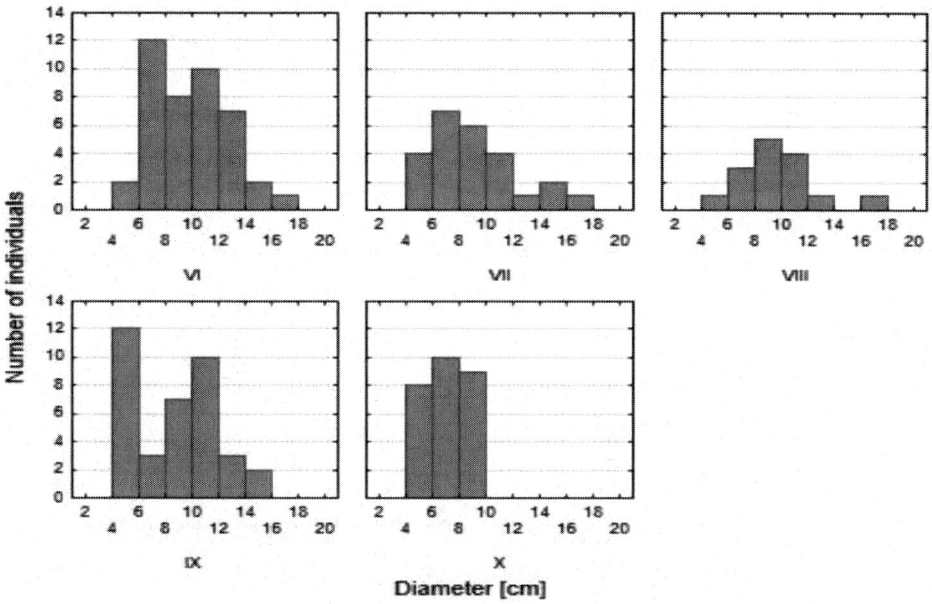


Figure 9. The diameter range values for *Populus tremula* on the plots of the stand in the unthinned forest.

Figure 10. Canopy cover in the unthinned forest (Photo by Katarzyna Żołna).

In the thinned forest individuals from the 0.5-2 m height class were the most frequent and had a mean density of 13,450 individuals per hectare (Table 2). The respective value for the up to 0.5 m height class was 9,025, whereas the density of the highest individuals, >2m high, was the lowest. The species richness within the three seedling height classes was also much larger in the up to 0.5 m and 0.5-2 m classes (12 and 13 species in the two classes mentioned, respectively) (Tables 3a, b, c) than in the remaining >2 m class (five species). In all of the plots that were studied *Betula pendula*, *Populus tremula* and *Salix aurita* were the dominant species in all three height classes (Figures 12-13; Tables 3a, b, c).

**Table 2. The mean total number of *Betula pendula*, *Populus tremula* and *Salix aurita* individuals in the tree height classes that were distinguished per 2x2m plots and their density per hectare in the thinned and unthinned forests that were studied (differences significant at $p < 0.05$; according to the Mann-Whitney U test)**

| Variable | Thinned forest | Unthinned forest | Thinned forest | Unthinned forest |
|---|---|---|---|---|
| | mean per 2 x 2 m plot | mean per 2 x 2 m plot | mean density per ha | mean density per ha |
| total number of individuals up to 0.5 m high | 3.61 | 2.30 | 9,025 | 5,750 |
| total number of individuals 0.5-2 m high | 5.38 | 1.43 | 13,450 | 3,580 |
| total number of individuals >2 m high | 0.38 | 4.79 | 960 | 11,980 |
| *Populus tremula* up to 0.5 m high | 2.61 | 1.77 | 6,525 | 4,425 |
| *Salix aurita* up to 0.5 m high | 2.09 | 1.30 | 5,225 | 3,250 |
| *Betula pendula* 0.5-2.0 m high | 3.10 | 2.14 | 7,750 | 5,350 |
| *Populus tremula* 0.5-2 m high | 2.96 | 1.63 | 7,400 | 4,075 |
| *Betula pendula* >2 m high | 1.31 | 4.08 | 3,275 | 10,200 |
| *Populus tremula* >2 m high | 1.14 | 1.89 | 2,850 | 4,725 |
| *Salix aurita* >2 m high | 1.08 | 2.66 | 2,700 | 6,650 |

Figure 11. Canopy cover in the forest with a juvenile stand thinning (Photo by Katarzyna Żołna).

Figure 12. Young, <0.5 m and 0.5-2 m height classes of a regenerating stand in the thinned forest that is dominated by *Betula pendula* individuals (Photo by Małgorzata Żaczek).

Figure 13. Young, <0.5 m and 0.5-2 m height classes of a regenerating stand of the thinned forest that is dominated by *Populus tremula* individuals (Photo by Małgorzata Żaczek).

In contrast to the forest that was subjected to thinning, the mean density of individuals >2 m high was much larger in the forest with the unthinned stand (11,980 individuals per hectare) than the respective values for the other two height classes (Table 2). Similar to the thinned forest the first two height classes were the most diverse as far as the species composition is concerned. Thirteen species were recorded in each of them. However, the number of species that were present in the >2 m height class was also large (11 species were noted there) (Tables 4a, b, c). *Betula pendula* was dominant in all of the classes. Its density was highest in the >2 m high individuals, but *Populus tremula* reached a similar density in the case of those up to 0.5 m high, at least on some study plots.

**Table 3. Species composition (number of individuals per hectare) of natural regeneration within the a) <0.5 m, b) 0.5-2 m and c) >2 m height classes of the thinned forest**

(a)

| Plot symbol | Betula pendula | Populus tremula | Salix aurita | Sorbus aucuparia | Quercus robur | Quercus petraea | Fagus sylvatica | Pinus sylvestris | Frangula alnus | Padus serotina | Picea abies | Carpinus betulus | Larix decidua |
|---|---|---|---|---|---|---|---|---|---|---|---|---|---|
| I_t | 3,150 | 3,450 | 50 | 250 | 850 | 50 | 50 | 100 | 50 | 0 | 0 | 50 | 0 |
| II_t | 2,300 | 5,300 | 1,050 | 300 | 500 | 50 | | 200 | 100 | 50 | 0 | 0 | 0 |
| III_t | 800 | 4,950 | 750 | 150 | 150 | 0 | 0 | 300 | 0 | 0 | 0 | 0 | 0 |
| IV_t | 2,750 | 3,900 | 1,150 | 150 | 750 | 50 | 150 | 50 | 0 | 0 | 0 | 0 | 0 |
| V_t | 2,550 | 4,050 | 2,950 | 300 | 600 | 50 | 200 | 200 | 200 | 0 | 50 | 0 | 0 |

(b)

| Plot symbol | Betula pendula | Populus tremula | Salix aurita | Sorbus aucuparia | Quercus robur | Quercus petraea | Fagus sylvatica | Pinus sylvestris | Frangula alnus | Padus serotina | Picea abies | Larix decidua |
|---|---|---|---|---|---|---|---|---|---|---|---|---|
| I_t | 6,400 | 4,050 | 500 | 200 | 150 | 0 | 100 | 0 | 0 | 0 | 0 | 0 |
| II_t | 6,150 | 5,450 | 1,400 | 100 | 50 | 100 | 0 | 50 | 200 | 0 | 0 | 0 |
| III_t | 3,800 | 5,700 | 1,100 | 100 | 50 | 50 | 50 | 0 | 0 | 50 | 0 | 0 |
| IV_t | 9,400 | 8,100 | 1,500 | 0 | 0 | 0 | 50 | 150 | 0 | 0 | 0 | 0 |
| V_t | 4,600 | 3,900 | 3,400 | 150 | 0 | 0 | 0 | 0 | 0 | 0 | 50 | 100 |

(c)

| Plot symbol | *Betula pendula* | *Populus tremula* | *Salix aurita* | *Quercus robur* | *Larix decidua* |
|---|---|---|---|---|---|
| I_t | 300 | 350 | 150 | 0 | 0 |
| II_t | 450 | 250 | 0 | 0 | 0 |
| III_t | 500 | 300 | 250 | 0 | 0 |
| IV_t | 950 | 250 | 50 | 0 | 0 |
| V_t | 550 | 100 | 250 | 50 | 50 |

**Table 4. Species composition (number of individuals per hectare) of natural regeneration within the a) <0.5 m, b) 0.5-2 m and c) >2 m height classes of the unthinned forest**

(a)

| Plot symbol | Betula pendula | Populus tremula | Salix aurita | Sorbus aucuparia | Quercus robur | Quercus petraea | Fagus sylvatica | Pinus sylvestris | Frangula alnus | Padus serotina | Picea abies | Larix decidua | Carpinus betulus |
|---|---|---|---|---|---|---|---|---|---|---|---|---|---|
| VI_n-t | 2,200 | 1,900 | 250 | 200 | 850 | 0 | 0 | 50 | 650 | 50 | 50 | 0 | 0 |
| VII_n-t | 1,600 | 650 | 300 | 950 | 1,400 | 50 | 200 | 0 | 1,050 | 50 | 0 | 100 | 50 |
| VIII_n-t | 1,150 | 1,400 | 400 | 750 | 1,000 | 550 | 0 | 50 | 800 | 0 | 0 | 100 | 0 |
| IX_n-t | 3,300 | 1,000 | 200 | 550 | 500 | 50 | 50 | 0 | 300 | 0 | 100 | 0 | 0 |
| X_n-t | 1,850 | 800 | 150 | 300 | 350 | 50 | 0 | 200 | 100 | 0 | 0 | 100 | 0 |

(b)

| Plot symbol | Betula pendula | Populus tremula | Salix aurita | Sorbus aucuparia | Quercus robur | Quercus petraea | Fagus sylvatica | Pinus sylvestris | Frangula alnus | Padus serotina | Larix decidua | Salix caprea | Picea abies |
|---|---|---|---|---|---|---|---|---|---|---|---|---|---|
| VI_n-t | 1,100 | 700 | 300 | 50 | 100 | 0 | 0 | 0 | 0 | 0 | 0 | 0 | 0 |
| VII_n-t | 950 | 0 | 250 | 150 | 250 | 0 | 300 | 0 | 0 | 0 | 0 | 50 | 0 |
| VIII_n-t | 1,800 | 750 | 500 | 450 | 300 | 200 | 50 | 50 | 100 | 50 | 100 | 100 | 0 |
| IX_n-t | 3,050 | 2,300 | 250 | 50 | 50 | 0 | 50 | 0 | 100 | 0 | 200 | 0 | 50 |
| X_n-t | 1,750 | 400 | 400 | 100 | 300 | 0 | 0 | 0 | 0 | 0 | 50 | 0 | 150 |

(c)

| Plot symbol | Betula pendula | Populus tremula | Salix aurita | Sorbus aucuparia | Quercus robur | Quercus petraea | Fagus sylvatica | Frangula alnus | Padus serotina | Larix decidua | Salix caprea |
|---|---|---|---|---|---|---|---|---|---|---|---|
| VI_n-t | 6,050 | 1,300 | 0 | 0 | 0 | 0 | 0 | 0 | 0 | 0 | 0 |
| VII_n-t | 5,350 | 850 | 1,700 | 50 | 700 | 50 | 50 | 0 | 0 | 50 | 100 |
| VIII_n-t | 7,150 | 1,350 | 1,750 | 0 | 800 | 350 | 50 | 50 | 0 | 500 | 100 |
| IX_n-t | 9,000 | 3,350 | 700 | 50 | 0 | 0 | 0 | 50 | 50 | 0 | 200 |
| X_n-t | 11,000 | 2,150 | 900 | 0 | 50 | 50 | 0 | 0 | 0 | 50 | 0 |

As the results presented above show, the natural regeneration of the stand was observed in both forest types despite their different history (juvenile management of the stand *vs.* no treatment). However, there were many significant, qualitative and quantitative differences in the dynamics of the regeneration process between the woods, as expressed by their species composition and the density of seedlings in different height classes. According to the results of the Mann-Whitney U test (Table 2), these differences were observed in the case of the density of the individuals in all the three height classes. Higher values of the density of younger individuals (the first two classes) and lower density of the older ones (the third class) were recorded in the thinned forest, whereas the relations were the opposite in the unthinned one. Similar tendencies were recorded in the case of the total number of species in the height classes that were distinguished (results not presented here). Other significant differences were observed in the density of the most frequent species, i.e., *Betula pendula*, *Populus tremula* and *Salix aurita*. Their density in the up to 0.5 m and 0.5-2 m height classes were larger in the case of the thinned forest, whereas the density of individuals > 2 m high was higher in the unthinned forest (Table 2).

## Herb Layer Composition

### *Microcommunities in the Thinned Forest*

The total number of species that were noted on the plots of the thinned forest was 88. Among them, *Calamagrostis epigejos*, *Deschampsia flexuosa*, *Juncus conglomeratus*, *Rubus fruticosus* and the seedlings of *Salix aurita* and *Pinus sylvestris* were the most constant, whereas *Calamagrostis villosa*, *Carex brizoides*, *Calamagrostis epigejos* and *Deschampsia flexuosa* reached the highest abundance (Appendix 1).

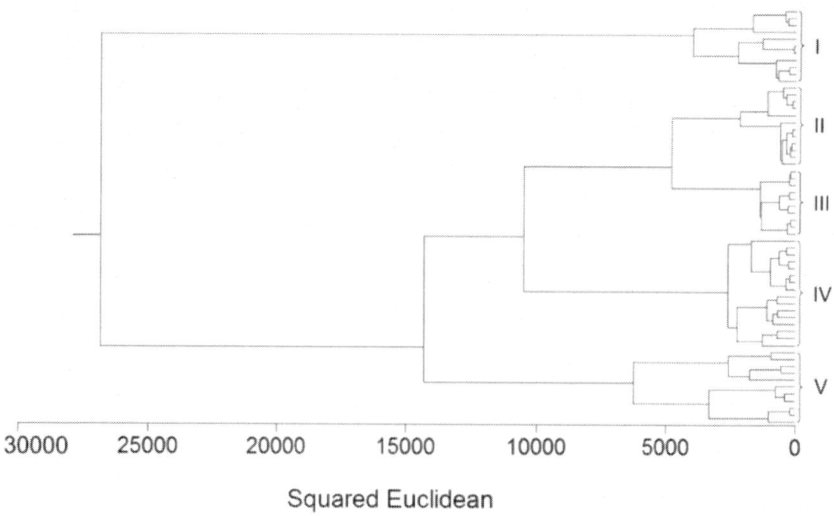


Figure 14. Dendrogram of the floristic similarity among the microcommunities that were distinguished in the herbaceous layer of the thinned silver birch forest. Names of the microcommunities are as follows: I – *Carex brizoides*, II – *Betula pendula–Calamagrostis epigejos*, III – *Calamagrostis epigejos–Holcus lanatus*, IV – *Calamagrostis villosa–Calamagrostis epigejos*, V – *Deschampsia flexuosa–Calamagrostis epigejos*.

Cluster analysis allowed five different microcommunities to be distinguished within the herbaceous layer of the thinned forest, which are named after their dominant species. A dendrogram of the floristic similarity of the microcommunities that were detected in the course of the analyses is illustrated on Figure 14.

In the first one, the *Carex brizoides* microcommunity (Figures 15-16), which was represented on 11 plots, the dominant species covered 30-80% of the herb layer and the most frequent accompanying species were usually grasses. This list included *Agrostis stolonifera, A. capillaris, Calamagrostis villosa, C. epigejos, Deschampsia caespitosa, D. flexuosa* and *Holcus lanatus.* The *Carex brizoides* microcommunity was distinctively different in its floristic composition from all of the other plots that had originated from the herb layer of the thinned forest (Figure 14). The Shannon diversity index of this microcommunity had the lowest value of all of the microcommunities that were distinguished (Table 5). These differences, however, were significant at $p = 0.0557$ according to the Kruskal-Wallis test. The mean cover of *Carex brizoides* on the plots of that microcommunity was 57.3%, whereas in the other four its cover ranged from 3.86% to 18.75% (Table 6).

The *Betula pendula-Calamagrostis epigejos* microcommunity (II) (Figure 17) was represented on 12 plots on which both dominant species were present. The *Betula pendula* cover ranged from 10% to 40%, whereas in the case of *Calamagrostis epigejos*, it varied between 5% and 50%. The most frequent species included *Agrostis stolonifera, A. capillaris, Calamagrostis villosa, Deschampsia flexuosa* and *Populus tremula.* The highest mean moss cover (exceeding 58%) and the second lowest tree and shrub cover among all five microcommunities of the thinned forest were recorded within the plots of the *Betula pendula–Calamagrostis epigejos* microcommunity (Table 5).

**Table 5. Mean (standard deviation) values of tree, shrub and moss layer cover, species richness and the Shannon diversity index in microcommunities of the thinned forest**

| Microcommunity (N) | Species richness Mean (SD) | Shannon diversity index Mean (SD) | Tree layer cover Mean (SD) | Shrub layer cover Mean (SD) | Herb layer cover Mean (SD) | Moss layer cover Mean (SD) |
|---|---|---|---|---|---|---|
| I (11) | 21.82 (4.60) | 2.24 (0.42) $^{b,c}$ | 74.09 (8.26) $^{b,c}$ | 14.36 (14.35) | 169.64 (59.31) $^{c}$ | 27.73 (23.49) $^{b}$ |
| II (12) | 21.58 (4.25) | 2.53 (0.20) $^{a,d,e}$ | 65.85 (9.02) $^{a,d,e}$ | 7.42 (7.44) | 131.00 (31.76) | 58.33 (24.80) $^{a,d}$ |
| III (10) | 20.50 (4.22) | 2.57 (0.24) $^{a,d,e}$ | 65.47 (6.26) $^{a,d,e}$ | 21.10 (10.85) $^{d}$ | 109.60 (16.62) $^{a,d,e}$ | 41.00 (22.83) |
| IV (16) | 19.50 (3.93) | 2.34 (0.27) $^{b,c}$ | 70.04 (9.69) $^{b,c}$ | 6.19 (6.68) $^{c}$ | 166.38 (32.02) $^{c}$ | 27.50 (25.56) $^{b}$ |
| V (11) | 22.73 (4.56) | 2.33 (0.16) $^{b,c}$ | 78.82 (11.51) $^{b,c}$ | 9.64 (15.34) | 189.64 (50.22) $^{c}$ | 40.00 (21.79) |

Explanations: symbols of microcommunities: I – *Carex brizoides*, II – *Betula pendula-Calamagrostis epigejos*, III – *Calamagrostis epigejos-Holcus lanatus*, IV – *Calamagrostis villosa-Calamagrostis epigejos*, V – *Deschampsia flexuosa-Calamagrostis epigejos*; different Arabic letters after the values indicate significant differences among the microcommunities according to the Kruskal-Wallis test ($p < 0.05$).

In the next (III) microcommunity – *Calamagrostis epigejos-Holcus lanatus* – the dominant species, which were recorded on all 10 plots, covered 5%-40% and 5%-30%, respectively (the mean cover of *Holcus lanatus* was the highest among all of the five microcommunities of the thinned forest) (Table 6). Similar to the previously mentioned microcommunity, the set of the most frequent accompanying species included *Agrostis stolonifera*, *A. capillaris*, *Calamagrostis villosa*, *Deschampsia caespitosa*, *Populus tremula* and also *Salix aurita* and *Betula pendula.* The total tree and herb layer cover had the lowest values whereas Shannon diversity index and the shrub cover were the highest among all five of the microcommunities of the thinned forest(Table 5).

**Table 6. Mean (standard deviation) values of the cover of the dominant herb layer species in the microcommunities of the thinned forest**

| Microcommunity (N) | Agrostis capillaris Mean (SD) | Calama-grostis villosa Mean (SD) | Calama-grostis epigejos Mean (SD) | Carex brizoides Mean (SD) | Desc-hampsia flexuosa Mean (SD) | Holcus lanatus Mean (SD) |
|---|---|---|---|---|---|---|
| I (11) | 11.10 (9.24) | 15.00 $(10.00)^{d}$ | 18.64 (14.68) | 57.27 $(17.37)^{b,c,d}$ | 10.56 $(8.82)^{e}$ | 6.67 $(2.58)^{c}$ |
| II (12) | 6.00 $(2.11)^{e}$ | 15.00 $(5.92)^{d}$ | 23.75 (13.67) | 3.86 $(1.95)^{a}$ | 7.82 $(7.77)^{e}$ | 3.67 $(2.31)^{c}$ |
| III (10) | 8.33 (5.00) | 15.00 $(8.66)^{d}$ | 16.00 $(11.50)^{e}$ | 6.50 $(4.36)^{a}$ | 5.13 $(2.42)^{e}$ | 14.00 $(8.10)^{a,b,d,e}$ |
| IV (16) | 18.18 (9.56) | 40.63 $(9.98)^{a,b,c,e}$ | 22.19 (10.48) | 12.00 $(10.17)^{a}$ | 13.75 $(9.04)^{e}$ | 5.56 $(1.67)^{c}$ |
| V (11) | 23.88 $(11.59)^{b}$ | 16.25 $(10.26)^{d}$ | 38.18 $(20.40)^{c}$ | 18.75 (12.17) | 44.00 $(18.97)^{a,b,c,d}$ | 5.00 $(0.00)^{c}$ |

Explanations: symbols of the microcommunities: I – *Carex brizoides*, II – *Betula pendula-Calamagrostis epigejos*, III – *Calamagrostis epigejos-Holcus lanatus*, IV – *Calamagrostis villosa-Calamagrostis epigejos*, V – *Deschampsia flexuosa-Calamagrostis epigejos*; different Arabic letters after the values indicate significant differences among the microcommunities according to the Kruskal-Wallis test ($p < 0.05$).

Figure 15. A thinned forest with the herb layer dominated by *Carex brizoides* (microcommunity I) (Photo by Anna Orczewska).

Figure 16. *Carex brizoides*, an expansive, native sedge species, which inhibits the establishment of typical woodland species (Photo by Anna Orczewska).

Figure 17. Thinned, silver birch forest with an herb layer that is representative of the *Betula pendula-Calamagrostis epigejos* microcommunity (Photo by Katarzyna Żołna).

The fourth microcommunity – *Calamagrostis villosa-Calamagrostis epigejos* – was most common in the thinned forest since it was distinguished on 16 plots in total. The two main dominants of its herb layer were recorded on all of the plots. The abundance of *Calamagrostis villosa* ranged from 30% to 60%, whereas in the case of *Calamagrostis epigejos*, it was lower (5% to 40%). The most common species were *Agrostis stolonifera*, *A. capillaris*, *Betula pendula* and *Deschampsia flexuosa.* The cover of the shrub and moss layers was the lowest among all of the microcommunities of the thinned forest (Table 5).

The abundance of the two main dominant species of the *Deschampsia flexuosa-Calamagrostis epigejos* microcommunity (V) varied from 10% to 70% and from 20% to 80%, respectively, and had the highest mean cover values among all of the microcommunities of the thinned forests, 44% and more than 38%, respectively (Table 6). In addition, *Carex brizoides*, *Agrostis capillaris* and *Calamagrostis villosa* were among the most frequent species in the herb layer of that microcommunity. The cover of the tree and herb layer and the mean species richness were the highest among all five of the microcommunities of the thinned forest (Table 5).

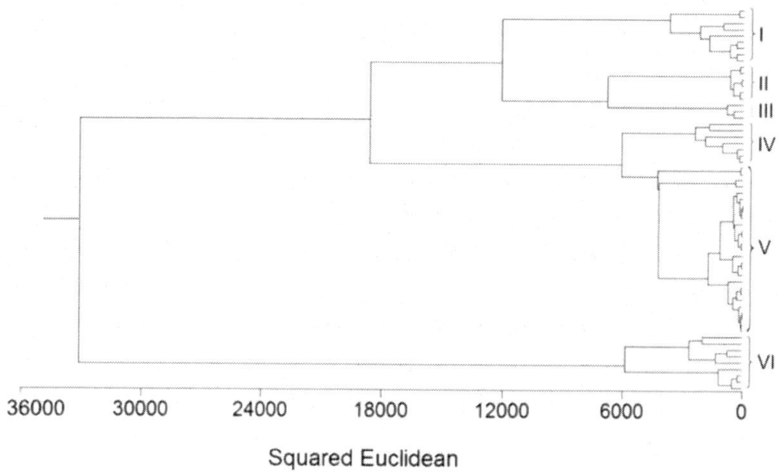


Figure 18. Dendrogram of the floristic similarity among the microcommunities that were distinguished in the herbaceous layer of the unthinned forest. Names of the microcommunities are as follows: I – *Carex brizoides-Calamagrostis villosa*, II – *Calamagrostis villosa*, III – *Calamagrostis villosa-Deschampsia flexuosa*, IV – *Agrostis capillaries-Calamagrostis epigejos*, V – *Agrostis capillaries-Calamagrostis villosa-Molinia caerulea-Deschampsia flexuosa-Vaccinium myrtillus*, VI – *Pteridium aquilinum*.

**Table 7. Mean (standard deviation) values of tree, shrub and moss layer cover, species richness and the Shannon diversity index in the microcommunities of the unthinned forest**

| Microcommunity (N) | Species richness Mean (SD) | Shannon diversity index Mean (SD) | Tree layer cover Mean (SD) | Shrub layer cover Mean (SD) | Herb layer cover Mean (SD) | Moss layer cover Mean (SD) |
|---|---|---|---|---|---|---|
| I (9) | 17.00 (2.69) [d] | 2.04 (0.25) [d] | 95.03 (7.20) [c,d] | 26.33 (10.16) | 156.22 (29.8) [e] | 19.38 (26.25) [e] |
| II (6) | 17.33 (1.97) | 1.89 (0.21) [d,e] | 90.12 (7.43) | 22.83 (9.37) | 115.00 (15.7) | 42.50 (29.28) |
| III (3) | 20.33 (3.21) | 2.22 (0.09) | 85.79 (11.46) [a] | 33.33 (10.41) | 201.33 (44.0) [e] | 35.00 (27.84) |
| IV (7) | 23.86 (4.38) [a] | 2.55 (0.24) [a,b,f] | 87.07 (10.62) [a] | 21.29 (9.69) | 146.71 (50.0) | 17.33 (35.65) [a,e] |
| V (27) | 18.37 (4.53) | 2.35 (0.26) [b] | 94.76 (7.63) | 28.33 (15.01) | 84.78 (32.7) [a,c,f] | 49.64 (29.12) [a,d] |
| VI (9) | 18.33 (3.57) | 1.99 (0.19) [d] | 91.80 (8.77) | 17.00 (13.06) | 192.00 (31.7) [e] | 22.50 (16.48) |

Explanations: symbols of the microcommunities: I – *Carex brizoides-Calamagrostis villosa*, II – *Calamagrostis villosa*, III – *Calamagrostis villosa-Deschampsia flexuosa*, IV – *Agrostis capillaris-Calamagrostis epigejos*, V – *Agrostis capillaris-Calamagrostis villosa-Molinia caerulea-Deschampsia flexuosa-Vaccinium myrtillus*, VI – *Pteridium aquilinum*; different Arabic letters after the values indicate significant differences among the microcommunities according to the Kruskal-Wallis test ($p < 0.05$).

***Microcommunities of the Unthinned Forest***

The total number of herbaceous vascular plants that were recorded on the plots of the forest with the unthinned stand was 78; *Calamagrostis epigejos*, *C. villosa*, *Trientalis europaea*, *Betula pendula* and *Rubus fruticosus* had the highest frequency/constancy (Appendix 2). The list of species with the highest mean cover in plots included *Pteridium aquilinum*, *Calamagrostis villosa*, *Carex brizoides*, *Deschampsia flexuosa* and *Molinia caerulea*.

Six microcommunities were distinguished within the herb layer of the unthinned forest in the course of cluster analyses. Their floristic similarity is presented in Figure 18 and the mean values of species richness (the number of species) and diversity (Shannon index), mean cover of trees, shrubs and mosses are given in Table 7.

Microcommunity I, *Carex brizoides-Calamagrostis villosa*, was dominated by *Carex brizoides*, for which the cover ranged from 30% to 80% (mean exceeded 51%); the highest in all of the six microcommunities that were distinguished – Table 8) and *Calamagrosits villosa*, which covered 10% to 70%. Both dominant species were present on each plot of this microcommunity. The tree canopy cover was the highest among all six of the microcommunities, whereas the species richness and diversity were among the lowest (Table 7).

Microcommunity II (Figure 19), which was represented by six plots, was dominated with *Calamagrostis villosa*, which had no less than 50% cover in every plot and its mean cover (55%) was the highest among all of the microcommunities of the unthinned woods (Table 8). *Carex brizoides* and *Molinia caerulea* were the most common accompanying species. The mean number of species and the Shannon diversity index for the herb layer were some of the lowest, whereas the moss cover was the second highest among the microcommunities of the unthinned forest (Table 7).

The third microcommunity – *Calamagrostis villosa-Deschampsia flexuosa* – covered a smaller proportion of the unthinned forest, since it was represented on only three plots. *Calamagrostis villosa* had cover ranging from 30% to 60%, whereas the second dominant species, *Deschampsia flexuosa*, was slightly less abundant, although it never reached less than 50% cover (mean for all of the plots was 56.67%; the highest among all of the microcommunities of the unthinned wood – Table 8). The most common accompanying species were *Calamagrostis epigejos*, *Molinia caerulea* and *Agrostis capillaris.* The tree layer cover of that microcommunity had the lowest value, while the shrub layer had one of the highest, which was similar to the mean values of species richness and diversity (Table 7).

Within the fourth microcommunity, *Agrostis capillaris-Calamagrostis epigejos*, the dominant species had 10% to 60% cover (mean per plot = 28,57%, the highest of all of the microcommunities – Table 8) and 1% to 20% cover, respectively, and were recorded in all seven plots that represented that microcommunity. *Calamagrostis villosa*, *Rubus fruticosus*, *Carex brizoides* and *Holcus mollis* were the most frequent and abundant accompanying species, although they were not always present on every plot. The cover of the tree, shrub and moss layers in the *Agrostis capillaris-Calamagrostis epigejos* microcommunity were among the lowest, whereas the herb layer cover, species richness and diversity had the highest values (Table 7).

**Table 8. Mean (standard deviation) values of the cover of the dominant herb layer species in the microcommunities of the unthinned forest**

| Microcommunity (N) | *Agrostis capillaris* Mean (SD) | *Calamagrostis villosa* Mean (SD) | *Calamagrostis epigejos* Mean (SD) | *Carex brizoides* Mean (SD) | *Deschampsia flexuosa* Mean (SD) | *Juncus conglomeratus* Mean (SD) |
|---|---|---|---|---|---|---|
| I (9) | 9.17 (5.85) | 35.00 (19.27) | 10.00 (4.63) | 51.11 (15.37) [c,d,f] | 10.00 (11.18) [c] | 1.80 (1.79) |
| II (6) | 7.50 (6.12) [d] | 55.00 (8.37) [d,e] | 6.67 (2.58) | 16.67 (5.77) | 7.50 (3.54) [c] | 3.00 (2.31) |
| III (3) | 15.00 (8.66) | 46.67 (15.28) [d,e] | 16.67 (5.77) | 7.50 (3.54)[a] | 56.67 (11.55) [a,b,d,e] | 3.00 (2.83) |
| IV (7) | 28.57 (16.76) [b] | 9.17 (5.85) [b,c] | 15.14 (8.38) | 13.33 (5.16) | 1.00 (0.00) [c] | 5.17 (2.86) [e] |
| V (27) | 10.24 (5.36) | 11.89 (7.30) [b,c] | 5.35 (2.73) | 8.46 (5.55) [a] | 11.64 (8.58) [c] | 1.33 (1.15) [d] |
| VI (9) | 15.83 (17.72) | 30.00 (15.00) | 7.22 (2.64) | 3.00 (2.83) [a] | 35.00 (23.75) | 2.00 (2.00) |

Explanations: symbols of the microcommunities: I – *Carex brizoides-Calamagrostis villosa*, II – Calamagrostis villosa, III – *Calamagrostis villosa-Deschampsia flexuosa*, IV – *Agrostis capillaris-Calamagrostis epigejos*, V – *Agrostis capillaris-Calamagrostis villosa-Molinia caerulea-Deschampsia flexuosa-Vaccinium myrtillus*, VI – *Pteridium aquilinum*; different Arabic letters after the values indicate significant differences among the microcommunities according to the Kruskal-Wallis test ($p < 0.05$).

Figure 19. Unthinned forest with an herb layer that is dominated by *Calamagrostis villosa* (microcommunity II) (Photo by Anna Orczewska).

Figure 20. Unthinned forest with an herb layer that is dominated by *Pteridium aquilinum* (microcommunity VI) (Photo by Anna Orczewska).

The most common microcommunity, represented by 27 plots and named *Agrostis capillaris-Calamagrostis villosa-Molinia caerulea-Deschampsia flexuosa-Vaccinium myrtillus* did not have any single, dominant species, but rather a combination of them that occurred with a high constancy but with a diverse abundance on the plots. Among them, *Agrostis capillaris* was noted on 21 plots and its cover ranged from 5% to 20%. *Calamagrostis villosa* occurred on 27 plots with a cover that varied from 1% to 30%. The successive dominants, *Molinia caerulea*, *Deschampsia flexuosa* and *Vaccinium myrtillus*, were recorded on 17, 22 and 19 plots, where their cover reached 1% to 50%, 1% to 30% and 1% to 50%, respectively. The tree and shrub layer cover in this microcommunity were the second highest, whereas the herb layer was poorly developed compared to the other microcommunities. In contrast, the moss layer cover and Shannon diversity index reached the highest and the second highest values among all six of the microcommunities of the unthinned forest (Table 7).

The last microcommunity of the unmanaged post-fire forest, the *Pteridium aquilinum* community, differed from the others in its species composition (Figure 18). These differences were mainly expressed by a high coverage of *Pteridium aquilinum* (Figure 20) compared to the other herb layer species that were recorded on the plots (its cover ranged from 40% to 80%). The coverage of the most common accompanying grass species, namely *Deschampsia flexuosa*, *Molinia caerulea*, *Calamagrostis villosa* and *Agrostis capillaris*, were usually lower than in the remaining five microcommunities that were described above. The Shannon diversity index was the second lowest and the shrub layer cover the lowest among all of the microcommunities of the unthinned wood (Table 7).

Detailed information regarding species richness and diversity, the total cover of tree, shrub, herb and moss layers in all of the microcommunities of the unthinned post-fire forests that were studied are included in Table 7, whereas the cover of dominant species is given in Table 8.

**Table 9. Ecological characteristics of the species in the herbaceous layer in the thinned and unthinned forests (differences according to the Mann-Whitney U test, at p < 0.05; ns – no significant difference)**

| Variable | | Thinned forest (N = 60) mean ±SD | Unthinned forest (N = 61) mean (±SD) |
|---|---|---|---|
| Raunkiaer life forms | | | |
| phanerophytes | frequency | 6.75 (±1.31) | 6.82 (±1.58) ns |
| | cover | 4.47 (±1.84) | 2.59 (±1.29) |
| chamaephytes | frequency | 0.80 (±0.82) ns | 0.72 (±0.84) |
| | cover | 3.69 (±2.84) | 7.77 (±10.44) |
| hemicryptophytes | frequency | 11.75 (±3.24) | 8.98 (±2.83) |
| | cover | 10.23 (±3.91) | 10.24 (±4.77) ns |
| geophytes | frequency | 1.32 (±0.77) | 1.92 (±0.84) |
| | cover | 3.54 (±3.18) | 7.62 (±11.99) ns |
| therophytes | frequency | 0.48 (±0.70) ns | 0.34 (±0.60) |
| | cover | 1.70 (±1.55) | 2.21 (±1.79) |
| Grime life strategies | | | |
| c | frequency | 10.57 (±1.43) ns | 10.21 (±1.85) |
| | cover | 7.86 (±2.55) | 6.34 (±3.48) |
| s | frequency | 0.97 (±0.61) | 1.41 (±0.67) |
| | cover | 2.92 (±1.93) ns | 2.39 (±1.71) |
| r | frequency | 1.00 (±0.00) | 0.00 (±0.00) |
| | cover | 1.00 (±0.00) ns | 0.00 (±0.00) |
| cs | frequency | 3.85 (±1.59) | 3.20 (±1.79) |
| | cover | 8.39 (±11.31) ns | 8.70 (±7.22) |
| cr | frequency | 0.33 (±0.63) ns | 0.33 (±0.60) |
| | cover | 1.00 (±0.00) | 1.91 (±1.53) |
| Forest and non-forest species | | | |
| forest | frequency | 10.28 (±2.58) | 10.48 (±2.28) ns |
| | cover | 7.03 (±3.12) ns | 6.17 (±3.02) |
| non-forest | frequency | 7.68 (±2.38) | 5.62 (±2.05) |
| | cover | 7.73 (±2.83) | 6.57 (±3.45) |
| Dispersal mode | | | |
| barochores | frequency | 1.90 (±0.84) | 2.51 (±0.72) |
| | cover | 2.39 (±1.50) ns | 2.06 (±1.31) |
| unwinged anemochores with small diaspores | frequency | 1.25 (±0.91) | 1.38 (±0.88) ns |
| | cover | 3.49 (±4.56) | 11.76 (±20.00) |
| anemochores with diaspores with pappus or wings | frequency | 10.75 (±2.14) ns | 8.93 (±1.74) |
| | cover | 10.07 (±3.82) | 8.54 (±3.81) |
| myrmecochores | frequency | 1.12 (±1.06) | 1.57 (±0.94) |
| | cover | 2.72 (±1.80) ns | 1.99 (±1.49) |

Original source: Studia i Materiały Centrum Edukacji Przyrodniczo-Leśnej, 12, 2010, 377-387. Wpływ czyszczeń na rozwój roślinności runa w spontanicznych odnowieniach brzozowych po pożarze, in Polish, with an English summary, Table 4, Orczewska A., Obidziński A., Żołna K. Used with the kind permission of the Centre for Nature and Forestry Education (Centrum Edukacji Przyrodniczo-Leśnej), Forest Faculty of Warsaw University of Life Sciences – SGGW).

## Ecological Characteristics of the Herb Layer in the Thinned and Unthinned Forests

The herb layer in forest that was originally subjected to thinning differed from those in the forest without any treatments in many other aspects. In both of the forests that were studied, non-woodland species were dominant in their herbaceous layers. Their share in the forest with the thinned stand was 50% whereas in the unthinned forest, it was 42.3%. In contrast, there were 25% of woodland species among the herb layer species in the thinned forest and 23% in the unthinned forest. Moreover, the mean total species richness (18.8 species in the unthinned and 21.1 in the thinned forest, respectively), the mean number (5.6 and 7.7) and mean cover (6.5% and 7.7%) of non-forest species and the mean Shannon diversity index (2.22 and 2.4) were significantly higher in the thinned forest than in the forest with the unmanaged stand. Furthermore, in the herb layer of the thinned forest, the higher cover of competitive species (c strategy), a higher number of competitors that tolerate stress (cs strategy type) and a higher number and abundance of anemochores with winged diaspores were recorded compared to the unthinned forest (Table 9). In contrast, the herb layer of the forest that was not subjected to any silvicultural management was characterized by a higher cover of chamaephytes, a higher number of geophytes, a higher number of stress-tolerant and competitive-ruderal species (s and cr strategies), a higher number of barochores and myrmecochores and a higher cover of anemochores with small diaspores (Orczewska et al. 2010) (Table 9).

**Table 10. Herb layer species that had an affinity to the thinned and unthinned forests as distinguished based on their frequencies (Fisher exact probability test results; significance level at $p < 0.05$)**

| Thinned forest | | Unthinned forest | |
|---|---|---|---|
| **Species** | **p level** | **Species** | **p level** |
| *Agrostis stolonifera* | 0.0000 | *Calamagrostis villosa* | 0.0001 |
| *Brachypodium sylvaticum* | 0.0006 | *Maianthemum bifolium* | 0.0079 |
| *Carex pallescens* | 0.0050 | *Molinia caerulea* | 0.0006 |
| *Chamaenerion angustifolium* | 0.0278 | *Mycelis muralis* | 0.0299 |
| **Thinned forest** | | **Unthinned forest** | |
| **Species** | **p level** | **Species** | **p level** |
| *Deschampsia flexuosa* | 0.0020 | *Pteridium aquilinum* | 0.0026 |
| *Hieracium lachenalii* | 0.0028 | *Viola mirabilis* | 0.0299 |
| *Holcus lanatus* | 0.0059 | *Betula pendula* | 0.0073 |
| *Holcus mollis* | 0.0243 | *Carpinus betulus* | 0.0159 |
| *Juncus conglomeratus* | 0.0000 | *Fagus sylvatica* | 0.0000 |
| *Lysimachia vulgaris* | 0.0000 | *Frangula alnus* | 0.0041 |
| *Stellaria longifolia* | 0.0290 | *Padus serotina* | 0.0159 |
| *Pinus sylvestris* | 0.0000 | *Salix caprea* | 0.0299 |
| *Salix aurita* | 0.0096 | | |
| *Sorbus aucuparia* | 0.0000 | | |

Original source: Studia i Materiały Centrum Edukacji Przyrodniczo-Leśnej, 12, 2010, 377-387. Wpływ czyszczeń na rozwój roślinności runa w spontanicznych odnowieniach brzozowych po pożarze, in Polish, with an English summary, Table 2, Orczewska A., Obidziński A., Żołna K. Used with the kind permission of the Centre for Nature and Forestry Education (Centrum Edukacji Przyrodniczo-Leśnej), Forest Faculty of Warsaw University of Life Sciences – SGGW).

**Table 11. The herbaceous species with a significantly different cover in the thinned and unthinned forests (results according to the Mann-Whitney U test)**

| Species | Thinned forest mean (±SD) | Unthinned forest mean (±SD) | p level |
|---|---|---|---|
| *Calamagrostis epigejos* | 23.75 (±15.63) | 8.14 (±5.45) | 0.000000 |
| *Holcus lanatus* | 8.367 (±6.29) | 5.37 (±7.04) | 0.015451 |
| *Juncus conglomeratus* | 3.85 (±2.44) | 2.48 (±2.29) | 0.016902 |
| *Maianthemum bifolium* | 3.00 (±2.11) | 1.52 (±1.37) | 0.024969 |
| *Molinia caerulea* | 5.25 (±4.57) | 16.37 (±14.47) | 0.000871 |
| *Stellaria media* | 1.00 (±0.00) | 3.50 (±3.02) | 0.012420 |
| *Betula pendula* | 15.00 (±9.27) | 3.30 (±2.39) | 0.000000 |
| *Populus tremula* | 4.64 (±2.07) | 2.42 (±1.93) | 0.000041 |
| *Salix aurita* | 7.46 (±5.11) | 2.56 (±1.97) | 0.000000 |

Original source: Studia i Materiały Centrum Edukacji Przyrodniczo-Leśnej, 12, 2010, 377-387. Wpływ czyszczeń na rozwój roślinności runa w spontanicznych odnowieniach brzozowych po pożarze, in Polish, with an English summary, Table 3, Orczewska A., Obidziński A., Żołna K. Used with the kind permission of the Centre for Nature and Forestry Education (Centrum Edukacji Przyrodniczo-Leśnej), Forest Faculty of Warsaw University of Life Sciences – SGGW).

**Table 12. Ecological characteristics of the herbaceous species in the microcommunities of the thinned forest (significant differences according to the Kruskal-Wallis test, at $p < 0.05$)**

| | Microcommunity (N) | | | | |
|---|---|---|---|---|---|
| Variable | I (11) | II (12) | III (10) | IV (16) | V (11) |
| c_Ph | 3.44 $^{b}$ | 5.86 $^{a}$ | 3.92 | 4.94 | 3.81 |
| c_H | 10.97 $^{c}$ | 7.26 $^{d,e}$ | 6.59 $^{a,d,e}$ | 12.47 $^{b,c}$ | 12.78 $^{b,c}$ |
| n_T | 0.73 | 0.42 | 0.20 $^{e}$ | 0.19 $^{e}$ | 1.00 $^{c,d}$ |
| c_c | 5.75 $^{d}$ | 7.85 | 6.39 $^{d}$ | 10.33 $^{a,c}$ | 7.75 |
| n_s | 0.64 $^{d}$ | 1.08 | 0.60 $^{d}$ | 1.25 $^{a,c}$ | 1.09 |
| c_cs | 4.34 $^{e}$ | 5.29 $^{e}$ | 3.59 $^{e}$ | 6.84 | 22.47 $^{a,b,c}$ |
| c_csr | 23.03 $^{b,c,d}$ | 4.84 $^{a}$ | 5.87 $^{a}$ | 7.91 $^{a}$ | 8.97 |
| c_For | 4.58 $^{d,e}$ | 6.59 | 4.18 $^{d,e}$ | 9.62 $^{a,c}$ | 8.80 $^{a,c}$ |
| n_NFor | 8.45 $^{d}$ | 7.50 | 8.80 $^{d}$ | 6.31 $^{a,c}$ | 8.09 |
| n_Auto | 1.18 $^{b,c,d}$ | 0.58 $^{a}$ | 0.50 $^{a}$ | 0.56 $^{a}$ | 0.82 |
| c_Auto | 52.09 $^{b,c,d}$ | 3.86 $^{a}$ | 6.50 $^{a}$ | 11.44 $^{a}$ | 17.56 |
| n_M | 1.09 $^{e}$ | 1.00 $^{e}$ | 0.80 $^{e}$ | 0.81 $^{e}$ | 2.00 $^{a,b,c,d}$ |
| c_An2 | 7.35 $^{d,e}$ | 8.59 $^{d}$ | 7.13 $^{d,e}$ | 12.82 $^{a,b,c}$ | 13.08 $^{a,c}$ |

Abbreviations: n_ – number (frequency) of species; c_ – cover of species; Ph – phanerophytes; H – hemicryptophytes; T – therophytes; c – competitive species; s – stress tolerant species; cs – competitive species tolerating stress; cr – competitive ruderals; csr – species tolerating moderate level of competition, stress and disturbance; For – forest species; NFor – non-forest species; Auto – autochores; M – myrmecochores; A2 – anemochores with diaspores with pappus and wings.

Symbols of the microcommunities: I – *Carex brizoides*, II – *Betula pendula-Calamagrostis epigejos*, III – *Calamagrostis epigejos-Holcus lanatus*, IV – *Calamagrostis villosa-Calamagrostis epigejos*, V – *Deschampsia flexuosa-Calamagrostis epigejos;* different Arabic letters after the values indicate significant differences among microcommunities.

Differences were also detected when the frequency and cover of individual herb layer species were compared between the forests with thinned and unthinned stands. It appeared that 14 species occurred significantly more often in the herb layer of the thinned forest, whereas in the case of the unthinned forest, there were 12 species that had an affinity to its herb layer (Table 10). Furthermore, there was a group of seven species with a distinctively higher abundance (cover) in the thinned forest and two for which the cover was higher in the unthinned wood (Table 11) (Orczewska et al. 2010).

**Table 13. Ecological characteristics of the herbaceous species in the microcommunities of the unthinned forest (significant differences according to the Kruskal-Wallis test, at $p < 0.05$)**

| | Microcommunity (N) | | | | | |
|---|---|---|---|---|---|---|
| **Variable** | **I (9)** | **II (6)** | **III (3)** | **IV (7)** | **V (27)** | **VI (9)** |
| c_Ph | 3.90 $^{b,e}$ | 2.00 $^{a}$ | 2.60 | 3.66 | 2.08 $^{a}$ | 2.33 |
| n_Ch | 0.11 $^{e,f}$ | 0.17 | 1.00 | 0.14 | 1.04 $^{a}$ | 1.11 $^{a}$ |
| c_H | 14.60 $^{e}$ | 11.45 | 18.24 $^{e}$ | 7.69 | 7.05 $^{a,c,f}$ | 13.97 $^{e}$ |
| c_G | 4.76 | 2.00 $^{f}$ | 2.78 | 4.76 | 2.31 $^{f}$ | 33.41 $^{b,e}$ |
| c_c | 8.17 $^{e}$ | 8.10 $^{e}$ | 8.60 $^{e}$ | 6.04 | 3.51 $^{a,b,c,f}$ | 11.33 $^{e}$ |
| n_s | 1.56 | 1.83 | 1.33 | 2.14 $^{e,f}$ | 1.19 $^{d}$ | 1.11 $^{d}$ |
| c_s | 3.00 | 2.00 | 2.33 | 3.10 | 1.64 $^{f}$ | 3.75 $^{e}$ |
| n_cs | 2.56 | 2.17 | 4.33 $^{d}$ | 1.71 $^{c}$ | 3.81 | 3.44 |
| c_cs | 5.85 | 4.31 | 20.67 $^{d}$ | 3.39 $^{c,f}$ | 8.09 | 15.82 $^{d}$ |
| n_cr | 0.78 $^{b,e}$ | 0.00 $^{a,d}$ | 0.33 | 0.86 $^{b,e}$ | 0.15 $^{a,d}$ | 0.22 |
| n_csr | 2.89 | 3.00 | 3.33 | 6.00 $^{e,f}$ | 2.44 $^{d}$ | 2.22 $^{d}$ |
| c_csr | 22.21 $^{b,e,f}$ | 6.82 $^{a}$ | 7.50 | 9.26 | 7.08 $^{a}$ | 6.63 $^{a}$ |
| c_For | 7.29 | 7.56 | 11.42 $^{d,e}$ | 4.40 $^{c,f}$ | 4.44 $^{c,f}$ | 8.94 $^{d,e}$ |
| c_NFor | 6.20 | 5.34 $^{c,d}$ | 9.30 $^{b,e}$ | 8.42 $^{b,e}$ | 5.82 $^{d}$ | 7.65 |
| n_Auto | 1.00 $^{f}$ | 0.67 | 1.00 $^{f}$ | 1.29 $^{f}$ | 0.59 | 0.22 $^{a,c,d}$ |
| c_Auto | 51.11 $^{c,e,f}$ | 12.75 | 6.25 $^{a}$ | 10.90 | 7.96 $^{a}$ | 3.00 $^{a}$ |
| c_Baro | 2.85 $^{e}$ | 2.17 | 2.33 | 2.95 $^{e}$ | 1.47 $^{a,d}$ | 2.19 |
| c_An1 | 6.17 $^{f}$ | 2.33 $^{f}$ | 5.00 $^{f}$ | 10.73 | 3.11 $^{f}$ | 47.63 $^{a,b,c,e}$ |
| n_An2 | 7.56 $^{c,d}$ | 8.33 | 11.00 $^{a}$ | 10.43 $^{a}$ | 8.96 | 8.78 |
| c_An2 | 9.34 | 10.78 $^{e}$ | 15.84 $^{e}$ | 7.23 | 6.12 $^{b,c,f}$ | 12.12 $^{e}$ |
| c_Endo | 5.37 $^{b}$ | 2.17 $^{a,d}$ | 3.67 | 5.60 $^{b}$ | 3.33 | 2.60 |

Abbreviations: n_ – number (frequency) of species; c_ – cover of species; Ph – phanerophytes; Ch- chamaephytes; H – hemicryptophytes; G – geophytes; c – competitive species; s – stress tolerant species; cs – competitive species tolerating stress; cr – competitive ruderals; csr – species tolerating moderate level of competition, stress and disturbance; For – forest species; NFor – non-forest species; Auto – autochores; Baro – barochores; An1 – unwinged anemochores with small diaspores; A2 – anemochores with diaspores with pappus and wings; Endo – endozoochores.

Symbols of the microcommunities: I – *Carex brizoides-Calamagrostis villosa*, II – *Calamagrostis villosa*, III – *Calamagrostis villosa-Deschampsia flexuosa*, IV – *Agrostis capillaris-Calamagrostis epigejos*, V – *Agrostis capillaris-Calamagrostis villosa-Molinia caerulea-Deschampsia flexuosa-Vaccinium myrtillus*, VI – *Pteridium aquilinum*; different Arabic letters after the values indicate significant differences among the microcommunities.

Differences were also observed in the number and cover of species that represent different life history traits among the microcommunities of the thinned and unthinned forests, and more differences were recorded in the case of the managed forest. For detailed results of the analyses of the number and cover of species with different ecological characteristics, which relate to life forms, life strategies and types of dispersal modes and association with forest or non-forest communities, in microcommunities of the thinned and unthinned woods, see Tables 12-13.

# Discussion

The discussion of the results that we obtained is rather difficult and incomplete since no studies were undertaken shortly after the fire. For these reasons, it is impossible to relate the state that we describe to any earlier stage of the secondary succession on these post-fire sites. Furthermore, what was pointed by Tyszkiewicz (1963), single-species silver birch stands do not occur naturally in Central Europe but can only develop on abandoned agricultural land, in places that have been prepared for Scots pine plantations and, eventually, on the sites of burned forests. Nevertheless, some important observations that we made are worth discussing and these results can be used for the future comparative studies of the forest dynamics of a post-fire origin.

## Stand Structure and Species Composition

Both stands started to develop spontaneously in the year following the fire, in spring of the 1993. The original massive appearance and successive growth of silver birch seedlings, which was reported by the forest managers, is not surprising. *Betula pendula* belongs to typical pioneer species (Białobok 1979). According to Whelan (1995), because it is an obligate seeder, it takes advantage of favorable post-fire conditions that are expressed with strong winds, which usually last until the second year after a fire, and its seeds are dispersed to burned site. For these reasons, it efficiently colonized the sites discussed here. As a result of the similar age of silver birch individuals, their rather uniform and monotonous height is not surprising. In the forest with the managed stand, the average tree height varied between 9 and 11 meters, whereas in the unthinned forest, silver birch individuals were slightly higher and reached 10-12 meters. These differences can be explained by the increased level of competition for light among the trees in the unmanaged stand (compared to the thinned forest). Similar height values were also recorded for European aspen, which only occasionally reached higher values that ranged from 10 to 13 meters. According to Obmiński (1977) and Jaworski (2011), both species grow very fast and reach their maximum heights at a young age – 15-20 years in the case of *Betula pendula* and 20-25 years for *Populus tremula*. After that period, the increase of their height is inhibited and is completed at the age of ca. 30-40 years. According to the standing crop quality classes tables (Czuraj 1990), which take into account tree age and height, the silver birch stand that was studied represents the first class of stand quality (Table 14). There is no doubt that the intensive height growth of silver birch tree individuals is related to the rich habitat of the fresh mixed broadleaved forest and fresh broadleaved forest site type that they occupy. According to Jaworski (2011), the optimal conditions for this species are on light soils with moderate moisture (it avoids both very dry and very damp sites) and fertility that develop on loamy sands.

Like their heights, the diameters of trees also change with a tree's age. Originally, at a young age, these increases are slow but they accelerate once a tree becomes older, usually before or just after the peak of an increase in height is achieved. Then, the tree diameter continues to increase even after its increase in height has stopped. Thus, tree diameter changes more dynamically than its height. This is especially easy to detect once stand management practices are implemented. Juvenile tree thinning and clearing leads to an increase in tree diameters in response to the availability of excessive light. In such a situation, a tree does not have to compete for light, which in turn allows it to invest in the growth of its diameter rather than in height (Obmiński 1977). Such a mechanism must be responsible for the larger mean diameters of the silver birch in the thinned forest compared to the unmanaged one (7.52 cm and 6.9 cm, respectively). Moreover, *Populus tremula* is a very frequent species in the forest with the unmanaged stand. Similar to silver birch, because it is a pioneer species, it also shows vigorous growth, especially on rich and moist soils (Spychalski 1974) such as the ones where these studies were conducted. The average diameters of 8.9 cm of *Populus tremula*, which are larger than in the case of silver birch on these sites, confirm the pattern that was observed by the above-mentioned authors.

**Table 14. Quality of the standing crop of silver birch (stand quality classes) (Czuraj 1990)**

| Stand age [years] | Stand quality classes and average tree height [m] | | | | |
|---|---|---|---|---|---|
| | I | II | III | IV | V |
| 10 | 5.7 | 4.3 | 3.4 | 2.2 | 1.5 |
| 15 | 9.6 | 7.5 | 5.6 | 4.4 | 3.3 |
| 20 | 11.3 | 9.5 | 7.7 | 6.2 | 4.8 |
| 28 | 14.0 | 12.0 | 9.8 | 7.8 | 6.2 |

Although both forests are characterized by a high density of tree individuals, the density was greater in the case of the thinned forest (2,260 individuals per hectare on average) than in the unthinned one (1,835 individuals per hectare, when the individuals of all of the tree species were taken into account). According to forest management guidelines, a birch stand at the age of 15 years that grows on a fresh broadleaved and fresh mixed-broadleaved forest site types should undergo the so-called early (juvenile) thinning, which would act in a similar way as naturally occurring self-thinning, with a goal of reducing the tree density to 300-400 individuals per hectare (Czuraj 1990; Jaworski 2011). At the age of 20-25 years, a further reduction of tree density should be implemented, thus leading to a density of 200-300 individuals per hectare, which should be achieved once a tree reaches its maturity and harvesting age (Szymański 1980). Such management practices are planned in the case of the silver birch stand that we studied.

As a result of thinning, the stand canopy cover in the managed forest is lower than in the unmanaged one (moderate in the thinned forest – 82.7% on average on the five study plots and full in the unthinned plots – 96.3%). Such a high tree canopy cover is typical for sapling stands (brush woods). However, the older the trees become the more cases of their natural deaths are observed (tree density declines as they grow in size). As a result of this natural process of self-thinning, the tree canopy cover is reduced and it becomes more open. Moreover, the growth processes on more fertile soils are more intensive and faster than on poorer sites (Szymański 1990).

Although the tree canopy cover in both stands was high (moderate and full), the crowns of birch and aspen were rather scarce with sparse foliage, thus allowing the light to penetrate to the forest floor and subsequently to develop an abundant herb layer. In the forest with the thinned stand where the tree cover was lower than in the forest that was not subjected to any management, the herb layer was more abundant (86.2% mean cover) than in the unmanaged one with a denser tree canopy (62.9% mean cover).

As a consequence of juvenile thinning, whose aim was to eliminate *Populus tremula*, the stand in the managed forest was exclusively composed of silver birch. The reason for the elimination of aspens during the process of early stand thinning is not fully understood since in the neighboring, unthinned stand, it reached larger diameters than the silver birch. Moreover, like *Betula pendula*, it belongs to the pioneer, forest species that play an important role in forest succession (Białobok 1973; Spychalski 1974). This is especially important in the secondary succession on post-fire, post-agricultural sites, on forest clearings and occasionally also on derelict, post-industrial sites (Obmiński 1977; Czarnowski 1978; Faliński 1998a; Gil et al. 1999). Although it is regarded as a forest species, European aspen is not associated with any particular forest site type (Bugała 1955). Nevertheless, it plays the role of a valuable pioneer crop species that facilitates the growth of the late-successional, shade-demanding trees by creating moderate shade conditions and producing the leaf litter of the intermediate biomass, which decomposes slightly faster than the one from silver birch leaves (Włoczewski 1968) (but note that according to Faliński 1998a, in the case of secondary succession in acidic and poor temperate zone habitats where naturally coniferous forests, especially those with Scots pine develop, aspen is regarded as an inhibitor of succession because of its litter). Furthermore, it is regarded as a valuable admixture tree species in a forest stand, due to its fast growth rate, especially on fertile and moist habitats as well as on fresh soils that have a texture ranging from loams to sands (loamy-sandy soils) (Jaworski 2011) such as those that we studied. It has been suggested that in a habitat of a mixed and fresh broadleaved forest site type, its share in the stand composition should not drop below 15% (Spychalski 1974). Presumably, the risk of a future infection of aspen by a fungus – *Melampsora populnea*, which is also potentially dangerous to Scots pine, which among others is a species that is planned to be planted here in later stages of succession, is the reason for the elimination of aspen from these sites. In addition, aspen avoids moist, acidic soils, especially those with deeper soil horizons for which pH drops below 4.5 (Jaworski 2011). Such conditions were observed in the research area (T. Wanic, personal communication) so perhaps this fact contributed to the decision to eliminate *Populus tremula* from the thinned forest.

The species composition of the unthinned stand was more diverse, but was dominated by *Betula pendula* and *Populus tremula*. These two light-demanding species have a wide ecological tolerance to climatic and edaphic conditions and have low demands for nutrients (Faliński 1998a). Nevertheless, they grow better in soils that have moderate fertility and moisture than on poor sites. According to Spychalski (1974), the density of natural seedlings of aspen is usually very high on burned sites. For these reasons, this author even suggests that a controlled fire be ignited on a site on which *Populus tremula* is planned to be sown afterwards. However, as was pointed out by Białobok (1973), aspen is a species that has a rather poor competitive ability due to the fact that it develops a shallow roots system, which allows it to take advantage of only the water and nutrients that accumulate in the upper layer of soil. Even under optimal site conditions, its growth rate decreases at an early age and the wood decay processes begin very quickly. The best fitness of *Populus tremula* in natural conditions is achieved on very fertile and moist sites that have compact soils that are weakly but permanently aggraded by surface runoff. Besides, as a highly light-demanding species, it dies very quickly in a stand with a fully closed canopy. The dominance of silver birch over aspen in the unthinned post-fire forest may also be the result of the fact that *Betula pendula* colonizes such sites more easily due to its massive seed germination, whereas *Populus tremula* develops more dynamically via offshoots (Spychalski 1974), which for obvious reasons was not the case here once the succession started. Furthermore, the saplings of aspen are even very sensitive to very short periods of water deficits, which contribute to their rapid death (Białobok 1973). Like the results that we obtained, the dominance of silver birch over aspen was also reported in the case of post-fire forests in the Rudy Raciborskie Forest District in southern Poland (Parusel 2006, Dobrowolska 2008) and in a habitat of a mixed, post-fire forest site type in southern Lithuania (Marozas et al. 2007). These results indicate that silver birch has a stronger competitive power than aspen.

Apart from *Betula pendula* and *Populus tremula*, which were the two main components of the stand of the unthinned forest, other species, namely *Quercus robur*, *Q. petraea*, *Pinus sylvetris*, *Larix decidua* and *Frangula alnus*, were recorded rather accidentally. All of these, except for *Frangula alnus*, are highly light-demanding, competitive trees. The higher density of *Q. robur* compared to the other tree species that were recorded is probably due to the presence of two old, unburned, individuals of oak in the neighborhood of the post-fire forest, which played the role of seed sources for the oaks that colonized the site. This was in contrast to the two other neighboring forest districts that were affected by fire at the same time as when the area that was studied was burnt, i.e., Rudy Raciborskie and Kędzierzyn, where mainly forests that had originally grown on poor sites were lost and successively spontaneously colonized by Scots pine (Dobrowolska 2008) such a situation was not recorded in the Rudziniec Forest District. Its rich habitats of fresh mixed broadleaved and fresh and moist broadleaved forest site types were colonized by silver birch instead, whereas Scots pine only played the role of an admixture in the unthinned forest. Such a distribution pattern of *Pinus sylvestris* might be the result of its competitive exclusion by broadleaved trees.

In conclusion, one can say that the tree species composition in the unthinned stand is typical for the young, early successional, post-fire forests that develop on a habitat of fresh mixed broadleaved and fresh and moist broadleaved forests site types in this part of Europe.

## Stand Regeneration Dynamics

Strong competition for space, nutrients, light and water between species with similar habitat requirements is a common phenomenon, and therefore it was easy to observe in the post-fire forests that were studied, especially in the one with unmanaged stand. The lack of any juvenile thinning resulted in a very high density of tree seedlings from the >2 m height class, which reached 11,980 individuals per hectare. Competition for light forces the trees to invest more into rapid growth on height, whereas their trunks do not increase in width proportionally to their height growth. For these reasons, the diameters of the trees in the unthinned forest were distinctively smaller than those in the stand of the managed forest. Moreover, the lack of a sufficient amount of light in the dense, unthinned wood suppressed many tree individuals and led to the successive death of many individuals (self-thinning and density-dependent mortality processes – Whelan 1995). Hanak (1994b) emphasizes that the changes that take place in a forest that is at the stage of a thicket, thus over the first thirteen years of its growth (13 years is the maximum age at which a stand is still regarded as a thicket) are very dynamic, because during that period usually ca. 99% of the initial number of individuals die. According to the report by Pietras-Bereżańska (1997), who investigated the post-fire woods in the neighboring Rudy Raciborskie, the initial number of tree seedlings of *Betula pendula* was ca. one million individuals per hectare. Thus, fifteen years later, 98.8% of seedlings had died, which confirms the rapid changes and intensive competition that took place on the sites that were studied.

For obvious reasons, that is juvenile thinning, older seedlings that belong to the >2 m height class are significantly less numerous in the thinned forest (on average 960 individuals per hectare). On the other hand, management practices allowed more light to penetrate the forest floor, which resulted in the greater number of seedlings from the <0.5 m and 0.5-2 m height classes compared to the forest with the unthinned stand. Such a dynamic reaction shows how great the potential of the site for forest regeneration is.

The lower number of seedlings of *Populus tremula* in the first two height classes of the unthinned forest that were studied compared to the managed wood might have been caused by the fact that aspen does not tolerate full canopy cover and dies when its seedlings are forced to grow in the deep shade. On the other hand, according to Białobok (1973) and Obmiński (1973), its saplings are more shade-tolerant and can grow in the shade that is created by silver birch. Such a pattern was observed in the wood that was studied since *Populus tremula* seedlings <0.5 m high were very abundant in both types of forests.

In all three height classes of the natural stand regeneration, there were thirteen species of trees and shrubs recorded in total. Despite this, the density of many of them in most plots did not exceed 200-250 individuals per hectare. These belong to the following group of species: *Fagus sylvatica*, *Frangula alnus*, *Padus serotina*, *Picea abies*, *Carpinus betulus*, *Salix caprea* and *Larix decidua*. Such observations are in accordance with the results that were obtained by Parusel (2006), who studied the spontaneous regeneration of a stand on a habitat of fresh mixed coniferous forest site type in the Rudy Raciborskie Forest District where he recorded seven species. Among them, *Betula pendula*, *Populus tremula*, *Salix aurita*, *S. caprea* and *Pinus sylvestris* were also listed. *Salix aurita* was recorded only sporadically there. In contrast, our results show that its density is among the highest and it is only lower than that of silver birch and aspen. *Salix aurita* is regarded as a Euro-Asian forest shrub species that prefers moist and boggy sites (Spychalski 1974), thus its abundant presence in the forests that we investigated may well be explained by its habitat preferences. Its densities in the >0.5 m and 0.5-2 m height classes were larger in the case of the thinned wood, whereas in the case of the <2 m height class the situation was reverse.

*Pinus sylvestris*, which was recorded so often by Parusel (2006) and Dobrowolska (2008) in the spontaneous post-fire forests in Rudy Raciborskie, where its density exceeded that reported for *Betula pendula* and *Populus tremula*, did not occur abundantly in any of the height classes that we distinguished (<300 individuals per hectare). Such a low proportion of Scots pine must be related with the fact that on more fertile sites such as the ones where we conducted the research, it is eliminated by silver birch and aspen.

From the two oak species that were present in the post-fire forests that we studied, *Quercus robur* is more effective in the colonization of these sites than *Quercus petraea*. This is probably due to the already mentioned presence of the two old individuals of pedunculate oak that survived a fire nearby and could potentially act as seed sources. Both oak species had higher densities in the unthinned forest although in the case of *Q. robur*, its regeneration in the thinned forest was also satisfactory. The appearance of oaks in the post-fire woods will facilitate the future management activities that will be designed to transform the single-species birch stands into multi-species ones, since the introduction of oak is planned on these sites. On the other hand, one may wonder how the oaks will grow in the unthinned forest, since, according to (Kolesničenko 1976), *Populus tremula* has a negative, allelopathic impact on oak seedlings. Moreover, for the future changes of the stand composition that are planned by the foresters, the presence of *Carpinus betulus* and *Fagus sylvatica*, which were recorded in the younger layers, is also very important. In addition, the appearance of these two shade-tolerant, climax species is a symptom of successional changes that take place slowly within the post-fires forests on these sites. However, these species are very prone to damage that is done by deer (Figure 21), and therefore would need special, active protection against herbivores at the young stage of their growth.

Figure 21. Thinned forest with young individuals of beech (*Fagus sylvatica*) that had been damaged by deer (Photo by Anna Orczewska).

## Herb Layer Composition and its Ecological Characteristics

### *Species Richness*

*Carex brizoides*, *Pteridium aquilinum* and different species of grasses, mainly *Calamagrostis epigejos*, *C. villosa*, *Deschampsia flexuosa* and *Molinia caerulea*, were those components that had the greatest impact on the physiognomy of the herbaceous layer in the post-fire woods that we investigated, regardless of the forest type. In total, 78 species were recorded in the herb layer of the unthinned forest, whereas there were 88 recorded in the thinned one. Among these, 66 species were present in both types of forest. Thus, the species richness of the post-fire forest that we studied was slightly higher than that one reported by Parusel (2006) for the neighboring region of Rudy Raciborskie, where in the period of 1995-

1997 he noted 60 in 1995 and 44 species in 1997. The mean number of species per plot in the same years was 17 and 14, respectively, whereas in the case of our studies, it was 19 in the unthinned and 21 in the thinned forest, and therefore these values were only slightly higher than the results of Parusel (2006). Like the number of species, the Shannon diversity indices for the plots in the thinned and unthinned forests that we investigated were also higher (2.4 and 2.2, respectively) than those calculated by Parusel (2006), which were 2.3 on the plots where the soil preparation that resulted in a mechanic disturbance was implemented and 1.7 on the plots that had no soil treatment after the fire. Despite these minor differences in species richness and diversity, which are probably caused by the differences in the habitats that are occupied by the forests in these two regions (less fertile in Rudy Raciborskie than in our study sites), there was a long list of species that were common to both areas, for example, *Agrostis capillaris*, *Calamagrostis epigejos*, *C. villosa*, *Chamaenerion angustifolium*, *Deschampsia flexuosa*, *Holcus lanatus*, *Rumex acetosella*, *Stellaria media*, *Taraxacum officinale* and many others. In general, species composition of the post-fire forests of both regions resembles other types of communities representing early stages of forest secondary succession, for example forest clearings. In the research of Markowski (1971, 1982), who made his investigations on former clearings, there was a long list of species which we also recorded in the course of our studies. Among others, this list involved: *Calamagrostis epigejos*, *Deschampsia caespitosa*, *D. flexuosa*, *Agrostis capillaris*, *Chamaenerion angustifolium*, *Rumex acetosella*, *Rubus idaeus*, *Maianthemum bifolium*, *Luzula pilosa*, *Vaccinium myrtillus*, *Cirsium arvense* and *Veronica officinalis*. In general, the species composition of the post-fire forests that were studied is the result of the processes of species migration and regeneration. Thus, it consists mainly of non-forest herbs, which are typical for the communities that develop on disturbed, ruderal sites, forest clearings, meadows and sandy areas, whereas the number of typical forest species is rather limited. The abundant occurrence of annuals and short-lived perennials, which are weedy species that quickly occupy any disturbed ground, including a site after burning, is typical for the early phases of regeneration of vegetation, which was emphasized by Whelan (1995) and Faliński (1998a). Such species composition is only temporary and it is quickly replaced by vigorously growing grasses, a phenomenon that was also observed in the case of sites that were studied soon after a fire. The high growth rate of grasses that sprout after a fire is well known and has been reported in many studies (Daubenmire 1968, Singh 1993; cited by Whelan 1995). Many grass species sprout either from their meristems, which are hidden at the leaf base and surrounded by thick tussocks (Whelan 1995), or from their underground parts, which are well protected in deeper layers of soil, and thus are good insulators. In the case of the site that was studied, *Calamagrostis epigejos* quickly increased its growth rate and subsequently dominated the landscape of the post-fire sites (K. Cymorek, personal communication). Its deeply located stolons, which reach up to 100 cm in length, must have allowed it to regenerate quickly and to conquer the vast areas that had been denuded by the fire due to the reduced competition from other species that had been killed in the fire. One may suspect that, apart from the *Calamagrostis epigejos*, only a few species are potentially able to survive a fire on a site (either their seeds, stolons or rhizomes). According to Parusel (2006), the presence of *Calamagrostis epigejos*, *C. villosa*, *Carex nigra*, *Deschampsia caespitosa*, *Juncus conglomeratus*, *J. effusus*, *Molinia caerulea*, *Pteridium aquilinum* and *Rubus plicatus* may be of such an origin (typical sprouters *sensu* Biswell 1974 and Keeley 1981) – whereas most species must have naturally colonized the herb layer of these forests afterwards. The high percentage share of species with a low constancy (accidental species) in the herb layer of the post-fire woods (83.5% in the unthinned and 96.6% in the thinned forest), was also observed by Parusel (2006), and this confirms that the species composition of such communities undergoes very dynamic transformations, which is a typical feature of young, early successional forests. On the other hand, most species that have the highest constancy on the plots are those that could have potentially survived the fire (for example, *Calamagrostis epigejos*, *C. villosa*, *Carex brizoides*, *Deschampsia flexuosa* and *Molinia caerulea*).

The massive occurrence of *Carex brizoides*, *Calamagrostis epigejos*, *Rubus fruticosus* and *Pteridium aquilinum*, which are competitive species that have low requirements for nutrients, a high resistance to harsh ecological conditions (with nutrient and water deficits), a high growth rate, especially in forests with an open canopy that shades other herbaceous species, the effective regeneration from stolons and rhizomes and high seed production and dispersal (Ouden et al. 2000; Griffiths and Filan 2007), poses a potential risk for the recovery process of the herb layer in the longer term in the post-fire woods that were studied. According to Falińska (1997), Goldberg and Werner (1983) and Dzwonko and Loster (1993), the high aboveground biomass that is produced by grasses and sedges creates very shady niches that have a dense tussock, which in turn makes the seed germination of other species, including trees and shrubs, very difficult. Thus, species with such characteristics, namely expansive and invasive ones, play the role of inhibitors of succession, *sensu* Connell and Slatyer (1977) and Faliński (1998b). Furthermore, their high aboveground biomass combined with low decomposition rates that are caused by a fire causes the biological activity of soil microorganisms to be very limited (Hauke-Pancewiczowa and Trzcińska 1980, Olszowska 2002; Zwoliński et al. 2004), which contributes to an increased risk of new fires being ignited in such circumstances (Dominik 1977; Szabla 1994). Thus, the canopy closure, which would consequently suppress the growth of the expansive, light-demanding species, like those listed above, could reduce this risk. However, stands that are composed of a single species with a translucent canopy such as silver birch cannot efficiently suppress the vigorous growth of grasses and ferns, including *Pteridium aquilinum*. Thus, changes in their composition should be considered by foresters in their management plans. Such stand species composition transformations are essential in order to reduce the risk of future fires of these forests. In addition, the highly competitive fern *Pteridium aquilinum*, which has been described as one of the five most common weed species in the world (Vetter 2009), which often appears after a fire (Matt et al. 2005), is a well-known inhibitor of the establishment of other forest herbs and trees because of its high production of biomass and rapid growth (Gliessman and Muller 1978; Dolling et al. 1994; Ouden et al. 2000). In addition the suppression is also due to its biochemical interactions with other species. Its allelopathic chemicals, which are present in the aboveground biomass and are leached and released during the decomposition, inhibit the growth of other species in their vicinity (Gliessman and Muller 1978; Dolling et al. 1994). For these reasons, its suppression would also be required. On the other hand, however, one may hope that the autotoxicity of bracken fern, which progresses with the growing age of this species, can help to reduce its abundance in the coming years, at least to some extent.

***Herb Layer Horizontal Structure***

The results of the cluster analyses that we conducted allowed us to distinguish many different microcommunities in both the thinned and unthinned forests that were studied. These results confirm the fact that the herb layer of these woods is not uniform but that it is rather a mosaic that is dominated by one or a few species. Such a diverse horizontal structure of the herb layer is probably related to the mechanic disturbance of the ground that took place while fighting the fire and during the successive removal of dead trees. Such observations were made by Markowski (1982), who recognized these activities as factors that very severely disturbed the structure of the vegetation that he studied. Thus, the microhabitats that were created during the disturbance could have influenced the current herb layer species composition and resulted in the differences among the microcommunities that were observed in the post-fire woods that were studied. An interesting aspect would be to study the temporal changes in the herb layer species composition of these sites. However, since no data are available on the community composition at the initial stage of the succession of these sites, such studies are not possible at this stage.

Despite the differences in the herb layer composition of the thinned and unthinned woods, as expressed in the diversity of their microcommunities, which is at least partly a consequence of the differences in the light conditions, there are also many similarities between them. A significant resemblance was observed in the case of the *Carex brizoides* microcommunity in the thinned forest and the *Carex brizoides-Calamagrostis villosa* community that developed within the unthinned woods. These similarities were mainly expressed in the lowest (or one of the lowest) values of the Shannon diversity indices and the highest cover of *Carex brizoides* compared with the other microcommunities that were distinguished. Thus, the inhibitive impact of *Carex brizoides* on other community components was confirmed by these low species diversity values.

Other similarities were observed between the *Deschampsia flexuosa-Calamagrostis epigejos* in the thinned forest and the *Calamagrostis villosa-Deschampsia flexuosa* microcommunity in the unthinned one. Apart from the similar dominant species, both microcommunities were characterized by a very high species richness and diversity compared to the other microcommunities of the thinned and unthinned forests.

### *Ecological Characteristics of the Herb Layer Species*

Despite the many obvious similarities in the species composition of their herbaceous layer between the forests, there were some differences recorded when some of the ecological characteristics traits of species were considered. The higher cover of chamaephytes and a higher number of geophytes (Figure 22) that were observed in the forest that did not undergo any thinning that created a higher number of therophytes in the thinned one could be explained by the lack of *vs.* the presence of silvicultural management, thus resulting in the differences in the light conditions and the level of the mechanic disturbance of the ground during the selective tree cutting. The latter promotes therophytes but suppresses chamaephytes.

Competition is the selective factor that played the most significant role in the herb layer composition of the woods that were studied, since competitive species (with c life strategy) reached the highest frequency and cover. However, their number and cover were higher in the thinned forest, whereas in contrast, stress-tolerant species (s strategy) had a higher frequency in the unthinned forest than in the thinned one. There is no doubt that the limited light availability in the unthinned forest was a selective pressure that caused stress to plants and its role was stronger in the unthinned forest than on the forest floor of the thinned wood.

Figure 22. *Oxalis acetosella*, a woodland, stress-tolerant species and a geophyte, that was often recorded in the herb layer of the unthinned post-fire wood (Photo by Anna Orczewska).

Figure 23. *Luzula pilosa*, a woodland species dispersed by ants, that was recorded in the herb layer of the post-fire wood with unthinned stand (Photo by Anna Orczewska).

The post-fire origin of the forests was also well reflected in the spectrum of the species that have different modes of seed dispersal, since the dominance of the anemochores was observed in their herb layer. Like zoochory, the dispersal of seeds *via* the wind is one of the most effective modes that is observed in nature. Due to the strong surface winds that occur after a fire, the dispersal distances of many species are enhanced. Thus, anemochory is favored in such post-fire advantageous conditions. This, in turn, results in the predominance of wind-dispersed species in the post-fire vegetation in fire-sensitive ecosystems (Whelan 1995). On the other hand, species that have a very limited dispersal ability, namely barochores and myrmecochores (Figure 23), are poorly represented on the sites that were studied although their frequency was higher in the unthinned forest than in the thinned one. Such relations may be explained by the fact that forest species (slightly better represented in the unthinned wood) are usually characterized by a low dispersal potential (Hermy et al. 1999, Dzwonko and Loster 2001), whereas many anemochores are the species that are typical for non-forest communities (significantly more frequent and abundant in the thinned wood in this study).

The herbaceous layer in both forests was dominated by non-woodland species, whereas herbs that are typically associated with forests were very poorly represented. Such proportions are not surprising for forests in the early successional stages of development. In this respect, to some extent, the herb layer composition of the post-fire woods resembled the one that was recorded in recent forests of a post-agricultural origin, where woodland species are also less frequent (observations reported by many authors, e. g., Peterken 1974; Dzwonko and Loster 1992; Matlack 1994, Brunet and von Oheimb 1998, Vellend 2003; Orczewska 2009b). Furthermore, similar mechanisms lay behind the low number and abundance of the woodland species that were present in the post-fire woods to those that act in the case of post-agricultural forests. Woodland herbs predominantly rely on vegetative propagation and produce a small number of seeds. Moreover, their seeds are usually short-lived and only have a longevity of up to five years – species that have no seed dormancy (Bierzychudek 1982, Bekker et al. 1998; Whigham 2004). Thus, in the case of post-fire woods where the viable, short-lived seeds of woodland species were burned during a fire, the recovery of typical woodland flora is very slow due to the lack of the propagules of these species that are left in the soil after a fire. The lack of their seeds combined with the already-mentioned poor dispersal capacities of many woodland species makes the colonization process very slow – the so-called dispersal limitation mechanism process, which has been mentioned in many studies (among others Peterken and Game 1984; Whitney and Foster 1988; Dzwonko 2001; Orczewska 2009a; 2010 – for their detailed review see Flinn and Vellend 2005). Nevertheless, the existence of a small patch of an unburned forest in the neighborhood of the unthinned forest may contribute to the better recovery of woodland species within its herb layer to some extent, in comparison with the forest floor in the thinned forest. In addition, its more shady conditions compared to the thinned wood might successively help to eliminate the vigorously growing grasses, thus acting as succession inhibitors, and subsequently, create good conditions for shade-tolerant forest species. Thus, one may expect that such circumstances will contribute to the faster recovery process of forest species in the herb layer of the unthinned wood.

### *Herb Layer Species with an Affinity to the Thinned and Unthinned Forests*

Both forest types had a group of species that showed an affinitiy to such woods; 14 species were recorded significantly more often in the thinned and 12 were asociated with the unthinned forest. Althouth it is impossible to relate these species groups to other studies, since there are no similar lists available for comparision, some conclusions can be drawn based on these results. The first observation is that in the case of the unthinned woods, there is a larger number of forest species than in the thinned wood. These differences might be related to the more shady conditions on the forest floor of the unthinned wood, which supresses the development of the herb layer and, in turn, provides safer conditions for shade-tolerant woodland species. The affinity of species such as *Maianthemum bifolium*, *Mycelis muralis*, *Viola mirabilis*, *Carpinus betulus*, *Calamagrostis villosa* and *Fagus sylvatica* to the herb layer of the forest that had no juvenile thinning of its stand fits such pattern. Furthermore, some of these species are stress-tolerant *sensu* Grime (1988). Since stress can be interpreted as poor light conditions in this particular case, the association of these species with the unthinned wood is fully understandable. Moreover, some of them are regarded as ancient woodland indicator species (*sensu* Hermy et al. 1999; Dzwonko and Loster 2001) and their presence in the unthinned wood can also be explained by the closer neighborhood of a small patch of the unburned wood to the unthinned than to the thinned forest.

Lastly, although we did not study the moss layer composition, it is worth mentioning that there were two light-demanding species that dominated that layer in these woods, namely *Polytrichum juniperinum* and *Ceratodon purpureus.* Their dominance was also recorded by Parusel (2006) in the post-fire forests of neighboring Rudy Raciborskie. According to Stefańska (2006), these two mosses are usually associated with poor, oligotrophic, coniferous forest sites. In addition, the development of an abundant moss layer is often recorded in post-fire forests. Such a phenomenon is often regarded as an inhibitor in the forest secondary succession (Whelan 1995).

## CONCLUSION

In the stands of both the thinned and unthinned post-fire forests, *Betula pendula* grew very vigorously. Its good condition was well expressed by its heights and diameters, which allow it to be classified as being representative of the first stand quality class. Juvenile thinning, whose goal was to eliminate *Populus tremula*, allowed the silver birch individuals to increase their diameters and contributed to an increase in the number of birch individuals with a width that exceeded 5 cm, compared to the silver birch individuals in the unthinned wood. Moreover, in the unthinned forest, where there was less space and light available, birch trees were thinner and grew higher than those in the thinned forest. In the stand of the unthinned forest, *Betula pendula* was more frequent than *Populus tremula*, whereas the latter had wider diameters and heights than silver birch.

Juvenile stand clearing contributed to the more dynamic regeneration of aspen; its individuals from the <0.5 m and 0.5-2 m height classes were more abundant there than in the unthinned forest and their density was the highest among all of the species that recorded. The species composition of shrubs and trees that regenerated after the fire was more diverse in the unthinned forest. Although a group of typical forest tree and shrub species was present in both forest types, their richness and density was higher in the forest with a stand that had not previously been subjected to any management. This is probably due to the denser canopy cover, which facilitates forest species and suppresses those that are typical for the earlier stages of forest succession. Although *Pinus sylvestris* was originally a dominant species of the stands prior to fire, its regeneration on these sites was very poor. In such rich habitats of mixed broadleaved and fresh and moist broadleaved forest site types, Scots pine is suppressed by more competitive, broadleaved species at this stage of a forest development mainly by *Betula pendula* and *Populus tremula*.

The dominant elements of the herb layer in the post-fire woods that undergo very dynamic changes, which are typical for early successional forests, are expansive grass and sedge species and bracken fern. Such a dominance pattern inhibits the growth of typical forest species (recruitment limitation mechanism) and contributes to the increased risk of future fires on these sites. Anemochory is the dominant form of seed dispersal among the herbaceous species that were present in post-fire woods that were studied. The original openness of these sites together with the strong winds that occurred there after the fire explain why wind-dispersed species took advantage and were the most successful in the colonization of these areas. Their appearance combined with the development of some sprouters, mostly grasses, contributed to the fast initiation of the recovery process of the community.

The increased amount of light that penetrated the forest floor of the thinned forest compared with the one without any juvenile thinning practices contributed to a qualitative and quantitative diversification of the herb layer composition of both forest types. The lack of human intervention in the structure and growth dynamics of the unthinned forest created more shady conditions within the forest floor. This, in turn, reduced the otherwise abundant growth of light-demanding species on one site and allowed some shade-tolerant, woodland herbs to develop on the other. The differences in the herb layer species composition between the thinned and unthinned forests were also expressed in the lists of species that showed an affinity to each of the forest types that were studied. In the case of the species that were associated with unthinned forest, there were many more typical woodland herbs than in the case of the thinned forest. In contrast, species that were typical for open types of vegetation communities were predominant in the latter one. The proportions of species that represented different life forms, life strategies, seed dispersal modes and that are typical for either forest or non-forest communities in the thinned and unthinned forests that we studied are similar to the differences that have often been reported in recent woods of a post-agricultural origin to some extent (in contrast to the so-called ancient forests = forests with long, undisturbed continuity). Thus, there are many similarities in the process of secondary succession between the post-fire woods and forests that develop on old, abandoned fields and meadows. The dispersal and recruitment limitation mechanisms work together in both types of ecosystems and shape the species composition of the post-fire plant communities that are present on these sites.

Despite the fact that there was a lack of similar studies undertaken in earlier stages of forest development on the sites that we selected, which made a comparison of our results impossible, they provide reasonably good and complex information about this stage of the community development. Furthermore, they can be a good reference for any future research that will be conducted on this post-fire site. This, in turn, would provide invaluable knowledge on the post-fire succession of fire-sensitive forest ecosystems. Since no extensive data on that process in such types of communities exist in the literature, continued research on these sites is very important. Moreover, the practical aspects of such investigations, which would set out the guidelines for forest managers, also need to be emphasized. The main conclusion that may be drawn from our studies on the stand structure of the woods is that thinning gives a potential to provide a better quality wood, and therefore that post-fire forests need management. However, when one looks at the forests that were studied from a perspective wider than just the production of a wood, the conclusions can be the opposite. Due to the fact the forests of that area are typical managed stands that are not protected by any forms of nature protection act, it would be unrealistic to expect that the managers will not focus on the timber production issues. Nevertheless, as the case of the unthinned forest shows, the exclusion of some areas from management offers the only opportunity to observe the process of successional changes that occur in nature without any human influence. Since, forest management practices attempt to copy the natural processes, it would be wise to allow nature to 'work' without any human intervention. This, in turn, would broaden our knowledge on the post-fire development of forest communities in this part of Europe. Moreover, our studies show that in the case of thinned forests, the recolonization of the herb layer by forest species is slower than in successional forests that lack any management treatments of their stands. Furthermore, thinning may contribute to an increased risk of fires on these sites by facilitating the growth of grasses in their herb layer. These conclusions are also important from the ecological point of view. If a forest is not exclusively understood as simply a stand and that all of its structural components are important, it is essential the recovery of its herbaceous layer be taken into account.

## ACKNOWLEDGMENTS

The authors would like to express their gratitude to the forest managers from the Rudziniec Forest District, Mr Hubert Wiśniewski and Mr Krzysztof Cymorek, for providing access to the former and current management plans regarding the study sites and to Mr Krzysztof Chojecki from the Katowice Regional Forest Direction and Mr Robert Pabian from the Rudy Raciborskie Forest District for providing access to the archival photographs that document the forest fire that occurred in 1992 in the area that was studied. We are also grateful to Mr Tomasz Wanic for preparing Figure 3.

## REFERENCES

[1] Bekker, R. M., Bekker, J. P., Grandin, U., Kalamees, R., Milberg, P, Poschlod, P., Thompson, K. and Willems, J. H. (1998). Seed size, shape and vertical distribution in the soil: indicators of seed longevity. *Functional Ecology*, *12*, 834-842.

[2] Bernadzki, E. and Brzeziecki, B. (1999). Wpływ metod odnowienia na różnorodność biologiczną lasów zagospodarowanych w Polsce. In K. Rykowski, G. Matuszewski and E. Lenart (Eds.), *Ocena wpływu praktyki leśnej na różnorodność biologiczną w lasach w Europie Środkowej. Studium w zakresie polskiej Ustawy o lasach i innych przepisów prawnych*, (pp. 21-38). Warszawa. Instytut Badawczy Leśnictwa.

[3] Białobok, S. (1973). *Topole, Populus L.* Warszawa, Poznań: Państwowe Wydawnictwo Naukowe.

[4] Białobok, S. (1979). *Brzozy, Betula L.* Warszawa-Poznań: Państwowe Wydawnictwo Naukowe.

[5] Bierzychudek, P. (1982). Life histories and demography of shade-tolerant temperate forest herbs: a review. *New Phytologist*, *90*, 757-776.

[6] Biswell, H. H. (1974). Effects of fire on chaparral. In T. T. Kozlowski and C. E. Ahlgren (Eds.), *Fire and Ecosystems*, (pp. 321-364). New York: Academic Press.

[7] Brown, J. K. and Smith, J. K. (Eds.) (2000). Wildland fire in ecosystems: effects of fire on flora. Gen. Tech. Rep. *RMRS-GTR-42-vol.2. Ogden, UT: U.S. Department of Agriculture, Forest Service, Rocky Mountain Research Station.*

[8] Brunet, J. and von Oheimb, G. (1998). Migration of vascular plants to secondary woodlands in southern Sweden. *Journal of Ecology*, *86*, 429-438.

[9] Budzáková, M., Galvánek, D., Littera, P. and Šibík, J. (2013). The wind and fire disturbance in Central European mountain spruce forests: the regeneration after four years. *Acta Societatis Botanicorum Poloniae*, *82*, 13-24.

[10] Bugała, W. (1955). Topole krajowe i obce, ich znaczenie gospodarcze. *Roczniki Sekcji Dendrologicznej Polskiego Towarzystwa Botanicznego, 10.*

[11] Certini, G. (2005). Effects of fire on properties of forest soils: a review. *Oecologia, 143,* 1-10.

[12] Ciepielowski, A. and Stolarek, A. (1995). Metodyka i wstępne wyniki badań zmian stosunków wodnych małych zlewni objętych pożarem lasu. *Sylwan, 11,* 101-109.

[13] Connell, J. H. and Slatyer, R. O. (1977). Mechanisms of succession in natural communities and their role in community stability and organization. *American Naturalist, 111,* 1119-1144.

[14] Czarnowski, M. S. (1978). *Zarys ekologii roślin lądowych.* Warszawa: Państwowe Wydawnictwo Naukowe.

[15] Czuraj, M. (1990). *Tablice zasobności i przyrostu drzewostanów: opracowanie na podstawie wyników pomiarowych badań polskich, niemieckich, radzieckich, holenderskich dla gatunków drzew leśnych naszej strefy geograficznej.* Warszawa: Państwowe Wydawnictwo Rolnicze i Leśne.

[16] Dobrowolska, D. (2008). Odnowienie naturalne na powierzchniach uszkodzonych przez pożar w Nadleśnictwie Rudy Raciborskie. *Leśne Prace Badawcze, 69,* 255-264.

[17] Dolling A., Zackrisson O. and Nilsson M. C. (1994). Seasonal variation in phytotoxicity of bracken (*Pteridium aquilinum* L. Kuhn). *Journal of Chemical Ecology, 20,* 3163-3172.

[18] Dominik, J. (1977). *Ochrona lasu. Podręcznik dla studentów wydziałów leśnych akademii rolniczych.* Wydanie I. Warszawa: Państwowe Wydawnictwo Rolnicze i Leśne.

[19] Dzwonko, Z. (2001). Migration of vascular plant species to a recent wood adjoining ancient woodland. *Acta Societatis Botanicorum Poloniae, 70,* 71-77.

[20] Dzwonko, Z. and Loster, S. (1992). Species richness and seed dispersal to secondary woods in southern Poland. *Journal of Biogeography, 19,* 195-204.

[21] Dzwonko, Z. and Loster, S. (1993). Relations between the floristic composition of isolated young woods and their proximity to ancient woodland. *Journal of Vegetation Science, 4,* 693-698.

[22] Dzwonko, Z. and Loster, S. (2001). Wskaźnikowe gatunki roślin starych lasów i ich znaczenie dla ochrony przyrody i kartografii roślinności. Typologia zbiorowisk i kartografia roślinności w Polsce. *Prace Geograficzne, 178,* 119-132.

[23] Falińska, K. (1997). *Ekologia roślin.* Wydanie II. Warszawa: Państwowe Wydawnictwo Naukowe.

[24] Faliński, J. B. (1998a). Deciduous woody pioneer species (*Juniperus communis, Populus tremula, Salix* sp. div.) in the secondary succession and regeneration. *Phytocoenosis N.S., 10,* 1-156.

[25] Faliński, J. B. (1998b). Invasive alien plants, vegetation dynamics and neophytism. *Phytocoenosis 10 (N.S.) Supplementum Carthographiae Geobotniacae, 9,* 163-187.

[26] Flinn, K. M. and Vellend, M. (2005). Recovery of forest plant communities in postagricultural landscapes. *Frontiers in Ecology and the Environment, 3,* 243-250.

[27] Fronczak, K. (2007). Ślady tamtych dni. *Echa Leśne, 8,* 9-12.

[28] Fronczak, K. (2012). *Jak Feniks z popiołów. 20 lat po wielkim pożarze.* Warszawa: Centrum Informacyjne Lasów Państwowych.

[29] Gil, W., Michalak, R. and Zachara, T. (1999). Wpływ nowych zalesień na różnorodność biologiczną na poziomie genetycznym, gatunkowym, ekosystemowym i krajobrazowym. In K. Rykowski, G. Matuszewski, and E. Lenart E (Eds.), *Ocena wpływu praktyki leśnej na różnorodność biologiczną w lasach w Europie Środkowej. Studium w zakresie polskiej Ustawy o lasach i innych przepisów prawnych*, (pp. 189-203), Warszawa: Instytut Badawczy Leśnictwa.

[30] Gliessman S. R. and and Muller C. H. (1978). The allelopathic mechanisms of dominance in bracken (*Pteridium aquilinum*) in Southern California. *Journal of Chemical Ecology, 4*, 337-362.

[31] Golberg, D. E. and Werner, P. A. (1983). Equivalence of competitors in plant communities: a null hypothesis from the Sierra Madre, Mexico, and a general model. *Ecology, 63*, 942-951.

[32] Griffiths, R. P. and Filan, T. (2007). Effects of bracken fern invasions on harvested site soils in Pacific Northwest (USA) coniferous forests. *Northwest Science, 81*, 191-198.

[33] Grime, J. P. (1988). The C-S-R model of primary plant strategies - origins, implications and tests. In L. D. Gottlieb and S. Jain (Eds.), *Evolutionary Plant Biology* (pp. 371-393). London: Chapman and Hall.

[34] Hanak, B. (1994a). Koncepcja zagospodarowania wielkich pożarzysk leśnych na terenie Regionalnej Dyrekcji Lasów Państwowych Katowice. *Sylwan, 138*, 85-93.

[35] Hanak, B. (1994b). Wróg czy sprzymierzeniec? Koncepcja hodowli i przebudowy samosiewnych populacji brzozy brodawkowatej na pożarzysku rudzko-rudziniecko-kędzierzyńskim roku 1992. *Trybuna Leśnika, 8*, 8-9.

[36] Hanak, B. (1997). Popożarowe refleksje nad zagospodarowaniem i skutkami wielkich pożarów leśnych w RDLP Katowice w roku 1992. In *Zagospodarowanie wielkich pożarzysk. Konferencja Naukowo-Techniczna, Jarnołtówek, 16-17 czerwca 1997*. ZG SITLiD i Zarząd Oddziału w Katowicach.

[37] Hauke-Pacewiczowa, T. and Trzcińska, M. (1980). Wpływ pożaru dna lasu na aktywność mikrobiologiczną gleby. *Roczniki Gleboznawcze, 31*, 33-41.

[38] Hawryś, Z. (1998). Odnowienie pożarzyska w lasach rudzko-rudziniecko-kędzierzyńskich. *Przyroda Górnego Śląska, 14*, 7.

[39] Hermy, M., Honnay, O., Firbank, L., Grashof-Bogdam, C. and Lawesson, J. E. (1999). An ecological comparison between ancient and other forest plant species of Europe, and the implications for forest conservation. *Biological Conservation, 91*, 9-22.

[40] Jaworski, A. (2011). Hodowla lasu. Tom 3. *Charakterystyka hodowlana drzew i krzewów leśnych*. Warszawa: Państwowe Wydawnictwo Rolnicze i Leśne.

[41] Jaworski, A. and Nosek, S. (1983). Badanie przydatności różnych metod określenia liczebności odnowienia naturalnego. *Sylwan, 2*, 31-41.

[42] Keeley, J. E. (1981). Reproductive cycles and fire regimes. In *Fire Regimes and Ecosystem Properties*. USDA Forest Service General Technical Report WO-26.

[43] Kleyer, M., Bekker, R. M., Knevel, I. C., Bakker, J. P, Thompson, K., Sonnenschein, M., Poschlod, P., Van Groenendael, J. M., Klimes, L., Klimesová, J., Klotz, S., Rusch, G. M., Hermy, M., Adriaens, D., Boedeltje, G., Bossuyt, B., Dannemann, A., Endels, P., Götzenberger, L., Hodgson, J. G., Jackel, A. K., Kühn, I., Kunzmann, D., Ozinga, W. A., Römermann, C., Stadler, M., Schlegelmilch, J., Steendam, H. J., Tackenberg, O., Wilmann, B., Cornelissen, J. H. C., Eriksson, O., Garnier, E. and Peco, B. (2008).

The LEDA Traitbase: A database of life-history traits of Northwest European flora. *Journal of Ecology*, *96*, 1266-1274.

[44] Kolesničenko, M. V. (1976). Biochimičeskie vzaimovlijanija drevesnych rasenij. *Moskva, Izd. Les. Prom.*

[45] Kwiatkowska-Falińska, A. (2008). Post-fire succession on abandoned fields in coniferous habitat (north-east Poland). *Acta Societalis Botanicorum Poloniae*, *77*, 235-254.

[46] Markowski, R. (1971). Regeneracja acidofilnych zbiorowisk leśnych na porębach wysp Wolina i południowo-wschodniego Uznamu. Poznańskie Towarzystwo Przyjaciół Nauk, Wydział Matematyczno-Przyrodniczy, *Prace Komisji Biologicznej*, *35*, 3-28.

[47] Markowski, R. (1982). Sukcesja wtórna roślinności na porębach lasów liściastych. Poznańskie Towarzystwo Przyjaciół Nauk, Wydział Matematyczno-Przyrodniczy, *Prace Komisji Biologicznej PWN*, *61*, 1-77.

[48] Marozas, V., Racinskas, J. and Bartkevicius, E. (2007). Dynamics of ground vegetation after surface fires in hemiboreal *Pinus sylvestris* forests. *Forest Ecology and Management*, *250*, 47-55.

[49] Matlack, G. R. (1994). Plant species migration in a mixed-history forest landscape in eastern North America. *Ecology*, *75*, 1491-1502.

[50] Matt S. McGlone M. S., Wilmshurst J. M. and Leach H. M. (2005). An ecological and historical review of bracken (Pteridium esculentum) in New Zealand, and its cultural significance. *New Zealand Journal of Ecology*, *29*, 165-184.

[51] Matuszkiewicz, W. (2001). *Przewodnik do oznaczania zbiorowisk roślinnych Polski.* Vademecum Geobotanicum. Warszawa: Państwowe Wydawnictwo Naukowe.

[52] Mirek, Z., Piękoś-Mirkowa, H., Zając, A. and Zając, M. (2002). *Flowering plants and pteridophytes of Poland, a checklist.* Kraków: Szafer Institute of Botany, Polish Academy of Sciences.

[53] Obmiński, Z. (1973). Ekologia osiki. In S. Białobok (Ed.), *Topole.* Warszawa: Państwowe Wydawnictwo Naukowe.

[54] Obmiński, Z. (1977). *Ekologia lasu.* Warszawa: Państwowe Wydawnictwo Naukowe.

[55] Olszowska, G. (2002). Wpływ pożaru w Nadleśnictwie Rudy Raciborskie na aktywność enzymatyczną gleb. *Roczniki Gleboznawcze*, *53*, 97-104.

[56] Orczewska, (2009a). Migration of herbaceous woodland flora into post-agricultural black alder woods planted on wet and fertile habitats in south western Poland. *Plant Ecology*, *204*, 83-96.

[57] Orczewska, A. (2009b). The impact of former agriculture on habitat conditions and distribution patterns of ancient woodland plant species in recent black alder (*Alnus glutinosa* (L.) Gaertn.) woods in southern Poland. *Forest Ecology and Management*, *258*, 794-803.

[58] Orczewska, A. (2010). Colonization capacity of herb woodland species in fertile, recent alder woods adjacent to ancient forest sites. *Polish Journal of Ecology*, *58*, 297-310.

[59] Orczewska, A., Obidziński, A. and Żołna, K. (2010). Wpływ czyszczeń na rozwój roślinności runa w spontanicznych odnowieniach brzozowych po pożarze. *Studia i Materiały Centrum Edukacji Przyrodniczo-Leśnej*, *12*, 377-387.

[60] Ouden, J., den, P. W. Hommel, F. M., Vogels, D., Janssen, B. and Smit, R. (2000). *The role of bracken (Pteridium aquilinum) in forest dynamics.* Wageningen Agricultural University, Wageningen, Netherlands.

[61] Parusel, J. B. (2006). Sukcesja roślinności na wielkim pożarzysku w Lasach Rudzkich. Environmental changes and biological assessment III: *Scripta Facultatis Rerum Naturalium Universitatis Ostraviensis*, *163*, 77-87.

[62] Peterken, G. F. (1974). A method for assessing woodland flora for conservation using indicator species. *Biological Conservation*, *6*, 239-245.

[63] Peterken, G. F. and Game, M. (1984). Historical factors affecting the number and distribution of vascular plant species in the woodlands of Central Lincolnshire. *Journal of Ecology*, *72*, 155-182.

[64] Pietras-Bereżańska, J. (1997). O wielkim pożarze lasu w Kuźni Raciborskiej z 1992 roku - w 5 lat później. *Wszechświat*, *10*, 259-261.

[65] *Plan Urządzania Gospodarstwa Leśnego Nadleśnictwa Rudziniec, Obręb Rachowice. Tom 2. Opis taksacyjny lasu. Okres gospodarczy, 1, 01, 1985 do 31.12.1994*. 1985. p. 593.

[66] Rozwałka, Z. (2003). *Zasady Hodowli Lasu*. Generalna Dyrekcja Lasów Państwowych.

[67] Spychalski, Z. (1974). *Hodowla lasu*. Tom 2. Warszawa: Państwowe Wydawnictwo Rolnicze i Leśne.

[68] Stefańska, E. (2006). Zmiany składu gatunkowego fitocenoz w przebiegu sukcesji wtórnej na siedlisku boru świeżego w borach dolnośląskich. *Badania Fizjograficzne nad Polską Zachodnią. Wydawnictwo Poznańskiego Towarzystwa Przyjaciół Nauk. seria B - Botanika*, *55*, 105-117.

[69] Szabla, K. (1994). Warunki powstawania i rozwoju pożarów, niektóre działania organizacyjne oraz aktualne zagadnienia hodowlane i ochronne na pożarzysku w Nadleśnictwie Rudy Raciborskie. *Sylwan*, *138*, 75-83.

[70] Szymański, S. (1980). Cele, metody i techniki trzebieży. In *Trzebieże*, Warszawa: Państwowe Wydawnictwo Naukowe.

[71] Szymański, (1990). *Ekologiczne podstawy hodowli lasu*. Warszawa: Państwowe Wydawnictwo Rolnicze i Leśne.

[72] Tyszkiewicz, S. (1963). Las jako przedmiot gospodarki. Ważniejsze gatunki drzew leśnych jako przedmiot uprawy. In S. Tyszkiewicz and Z. Obmiński (Eds.). *Hodowla i uprawa lasu*, Warszawa: Państwowe Wydawnictwo Rolnicze I Leśne.

[73] Vellend, M. (2003). Habitat loss inhibits recovery of plant diversity as forests regrow. *Ecology*, *84*, 1158-1164.

[74] Vetter, J. (2009). A biological hazard of our age: bracken fern [*Pteridium aquilinum* (L.) Kuhn] – a review. *Acta Veterinaria Hungarica*, *57*, 183-196.

[75] Whelan, R. J. (1995). *The ecology of fire*. New York, Cambridge University Press.

[76] Whigham, D. F. (2004). Ecology of woodland herbs in temperate deciduous forests. *Annual Review of Ecology, Evolution and Systematics*, *35*, 583-621.

[77] Whitney, G. C. and Foster, D. R. (1988). Overstorey composition and age as determinants of the understorey flora of woods of central New England. *Journal of Ecology*, *76*, 867-876.

[78] Włoczewski, T. (1968). *Ogólna hodowla lasu*. Warszawa: Państwowe Wydawnictwo Rolnicze i Leśne.

[79] Zwoliński, J., Matuszczyk, I. and Hawryś, Z. (2004). Właściwości chemiczne gleb i igieł sosny oraz aktywność mikrobiologiczna gleb na terenie pożarzysk leśnych z 1992 roku w Nadleśnictwach Rudy Raciborskie i Potrzebowice. *Leśne Prace Badawcze*, *1*, 119-133.

# Appendix 1. Mean, Maximum and Minimum Values of Cover and Constancy of Herbaceous Species in the Thinned Forest

| Species | Microcommunity 1 | | | Microcommunity 2 | | | Microcommunity 3 | | | Microcommunity 4 | | | Microcommunity 5 | | | Constancy |
|---|---|---|---|---|---|---|---|---|---|---|---|---|---|---|---|---|
| | mean | max | min | mean | max | min | mean | max | min | mean | max | min | mean | max | min | |
| *Abies alba* | 1 | 1 | 1 | | | | | | | | | | | | | I |
| *Achillea millefolium* | | | | 1 | 1 | 1 | 3 | 5 | 1 | | | | | | | I |
| *Agrostis stolonifera* | 15.8 | 30 | 5 | 6.8 | 10 | 5 | 8.2 | 20 | 5 | 15 | 30 | 5 | 10 | 20 | 5 | III |
| *Agrostis capillaris* | 11.1 | 30 | 1 | 6 | 10 | 5 | 8.3 | 20 | 5 | 18.2 | 30 | 5 | 23.9 | 40 | 1 | IV |
| *Alnus glutinosa* | | | | 1 | 1 | 1 | | | | | | | | | | I |
| *Anthoxanthum odoratum* | | | | 3 | 5 | 1 | | | | 5 | 5 | 5 | | | | I |
| *Athyrium filix-femina* | 5 | 5 | 5 | 2.3 | 5 | 1 | 5 | 5 | 5 | 2.3 | 5 | 1 | 5 | 5 | 5 | II |
| *Betula pendula* | 8.3 | 20 | 5 | 22.5 | 40 | 10 | 10 | 10 | 10 | 13.8 | 30 | 5 | 11.3 | 20 | 5 | IV |
| *Bidens frondosa* | | | | | | | | | | | | | 1 | 1 | 1 | I |
| *Brachypodium sylvaticum* | | | | 9 | 20 | 5 | 5 | 5 | 5 | 8.3 | 10 | 5 | 10 | 10 | 10 | I |
| *Calamagrostis villosa* | 15 | 30 | 5 | 15 | 20 | 5 | 15 | 20 | 5 | 40.6 | 60 | 30 | 16.3 | 30 | 5 | IV |
| *Calamagrostis epigejos* | 18.6 | 40 | 5 | 23.7 | 50 | 5 | 16 | 40 | 5 | 22.2 | 40 | 5 | 38.2 | 80 | 20 | V |
| *Calluna vulgaris* | 5 | 5 | 5 | 3 | 5 | 1 | 5 | 5 | 5 | 5.2 | 10 | 1 | 5 | 5 | 5 | I |
| *Carex brizoides* | 57.3 | 80 | 30 | 3.8 | 5 | 1 | 6.5 | 10 | 1 | 12 | 30 | 1 | 18.8 | 40 | 5 | IV |
| *Carex pallescens* | 1 | 1 | 1 | 3 | 5 | 1 | 5 | 5 | 5 | 2.3 | 5 | 1 | 2 | 5 | 1 | II |
| *Carex vulpina* | | | | | | | | | | 1 | 1 | 1 | | | | I |
| *Carpinus betulus* | | | | | | | | | | | | | 1 | 1 | 1 | I |
| *Cerasus avium* | | | | | | | | | | | | | 1 | 1 | 1 | I |
| *Chamaenerion angustifolium* | 1 | 1 | 1 | 1.8 | 5 | 1 | 2.3 | 5 | 1 | 1.8 | 5 | 1 | 2.3 | 5 | 1 | III |
| *Chenopodium album* | | | | 1 | 1 | 1 | | | | | | | | | | I |
| *Cirsium arvense* | | | | | | | 1 | 1 | 1 | | | | | | | I |

| Species | Microcommunity 1 | | | Microcommunity 2 | | | Microcommunity 3 | | | Microcommunity 4 | | | Microcommunity 5 | | | Constancy |
|---|---|---|---|---|---|---|---|---|---|---|---|---|---|---|---|---|
| | mean | max | min | mean | max | min | mean | max | min | mean | max | min | mean | max | min | |
| *Cirsium vulgare* | | | | | | | 1 | 1 | 1 | | | | | | | I |
| *Deschampsia caespitosa* | 8. 7 | 20 | 1 | 2.3 | 5 | 1 | 8 | 20 | 5 | 3 | 5 | 1 | 1.8 | 5 | 1 | II |
| *Deschampsia flexuosa* | 10. 6 | 30 | 5 | 7.8 | 30 | 1 | 5.2 | 10 | 1 | 13.7 | 30 | 5 | 44 | 70 | 10 | V |
| *Dryopteris carthusiana* | 2. 2 | 5 | 1 | 2.7 | 5 | 1 | 2 | 5 | 1 | 2.3 | 5 | 1 | 1 | 1 | 1 | III |
| *Dryopteris filix-mas* | 1 | 1 | 1 | | | | 5 | 5 | 5 | 1 | 1 | 1 | | | | I |
| *Epilobium ciliatum* | 1 | 1 | 1 | | | | 1 | 1 | 1 | | | | 1 | 1 | 1 | I |
| *Eupatorium cannabinum* | | | | 5 | 5 | 5 | | | | | | | | | | I |
| *Fagus sylvatica* | 3 | 5 | 1 | 1 | 1 | 1 | 1 | 1 | 1 | 3 | 5 | 1 | | | | I |
| *Festuca gigantea* | | | | 1 | 1 | 1 | 1 | 1 | 1 | | | | | | | I |
| *Frangula alnus* | 1 | 1 | 1 | 1 | 1 | 1 | 3 | 5 | 1 | 2 | 5 | 1 | 1.8 | 5 | 1 | II |
| *Galeopsis tetrahit* | 1 | 1 | 1 | 1 | 1 | 1 | | | | 1 | 1 | 1 | 1 | 1 | 1 | I |
| *Galium aparine* | | | | | | | | | | | | | 1 | 1 | 1 | I |
| *Hieracium* sp. | | | | | | | | | | 1 | 1 | 1 | | | | I |
| *Hieracium lachenalii* | | | | | | | 1 | 1 | 1 | 3 | 5 | 1 | 3 | 5 | 1 | I |
| *Hieracium pilosella* | 1 | 1 | 1 | | | | 1 | 1 | 1 | 3 | 5 | 1 | 1 | 1 | 1 | I |
| *Holcus lanatus* | 6. 7 | 10 | 5 | 3. 7 | 5 | 1 | 14 | 30 | 5 | 5. 6 | 10 | 5 | 5 | 5 | 5 | III |
| *Holcus mollis* | 15 | 30 | 5 | 7.5 | 10 | 5 | 5 | 5 | 5 | 10.6 | 30 | 5 | 13.8 | 30 | 5 | III |
| *Juncus conglomeratus* | 3. 7 | 5 | 1 | 3.9 | 5 | 1 | 4.7 | 10 | 1 | 4.5 | 10 | 1 | 2.2 | 5 | 1 | V |
| *Juncus effusus* | 1 | 1 | 1 | | | | | | | | | | | | | I |
| *Juncus tenuis* | | | | 5 | 5 | 5 | | | | 1 | 1 | 1 | 1 | 1 | 1 | I |
| *Larix decidua* | 1 | 1 | 1 | 1 | 1 | 1 | | | | | | | | | | I |
| *Leontodon autumnalis* | | | | 1 | 1 | 1 | | | | | | | 1 | 1 | 1 | I |
| *Lotus uliginosus* | | | | | | | 5 | 5 | 5 | 1 | 1 | 1 | 1 | 1 | 1 | I |
| *Luzula multiflora* | 3 | 5 | 1 | 5 | 5 | 5 | 1 | 1 | 1 | 4.2 | 5 | 1 | 1 | 1 | 1 | II |
| *Luzula pilosa* | 1.8 | 5 | 1 | | | | 1 | 1 | 1 | 1 | 1 | 1 | 3.7 | 5 | 1 | II |

## Appendix 1. (Continued)

| Species | Microcommunity 1 | | | Microcommunity 2 | | | Microcommunity 3 | | | Microcommunity 4 | | | Microcommunity 5 | | | Constancy |
|---|---|---|---|---|---|---|---|---|---|---|---|---|---|---|---|---|
| | mean | max | min | mean | max | min | mean | max | min | mean | max | min | mean | max | min | |
| *Lycopodium clavatum* | | | | 3 | 5 | 1 | | | | 2.3 | 5 | 1 | 1 | 1 | 1 | I |
| *Lycopus europaeus* | | | | | | | 1 | 1 | 1 | | | | | | | I |
| *Lysimachia vulgaris* | 3. 7 | 5 | 1 | 2.3 | 5 | 1 | 3.2 | 5 | 1 | 2.7 | 5 | 1 | 2.7 | 5 | 1 | IV |
| *Maianthemum bifolium* | | | | 3. 7 | 5 | 1 | | | | 2.6 | 5 | 1 | 3 | 5 | 1 | I |
| *Melampyrum pratense* | 5 | 5 | 5 | | | | 1 | 1 | 1 | 1 | 1 | 1 | 3.4 | 5 | 1 | I |
| *Moehringia trinervia* | 2.3 | 5 | 1 | 5 | 5 | 5 | 1 | 1 | 1 | | | | 1.8 | 5 | 1 | I |
| *Molinia caerulea* | 3 | 5 | 1 | 12.5 | 20 | 5 | 4.2 | 5 | 1 | 3. 7 | 5 | 1 | 7.5 | 10 | 5 | II |
| *Oxalis acetosella* | 3 | 5 | 1 | | | | | | | | | | | | | I |
| *Padus serotina* | | | | | | | | | | 1 | 1 | 1 | 1 | 1 | 1 | I |
| *Picea abies* | 2.3 | 5 | 1 | 1 | 1 | 1 | 1 | 1 | 1 | 1 | 1 | 1 | | | | I |
| *Pinus sylvestris* | 1 | 1 | 1 | 1 | 1 | 1 | 1 | 1 | 1 | 1 | 1 | 1 | 1 | 1 | 1 | V |
| *Poa nemoralis* | | | | | | | | | | 5 | 5 | 5 | | | | I |
| *Poa pratensis* | 5 | 5 | 5 | 3 | 5 | 1 | | | | | 0 | 0 | 3 | 5 | 1 | I |
| *Polygonum minus* | | | | | | | | | | | 0 | 0 | 1 | 1 | 1 | I |
| *Populus tremula* | 3. 7 | 5 | 1 | 6.3 | 10 | 5 | 5 | 5 | 5 | 3.9 | 5 | 1 | 4.1 | 5 | 1 | IV |
| *Potentilla reptans* | 5 | 5 | 5 | | | | 1 | 1 | 1 | | | | | | | I |
| *Pteridium aquilinum* | | | | 5 | 5 | 5 | | | | 20 | 20 | 20 | 11. 7 | 20 | 5 | I |
| *Quercus robur* | 1 | 1 | 1 | 1 | 1 | 1 | 2.7 | 5 | 1 | 1 | 1 | 1 | 2.3 | 5 | 1 | III |
| *Quercus rubra* | 1 | 1 | 1 | 1 | 1 | 1 | 1 | 1 | 1 | 1 | 1 | 1 | 1 | 1 | 1 | I |
| *Quercus petraea* | 2.3 | 5 | 1 | 1 | 1 | 1 | 3 | 5 | 1 | 1.6 | 5 | 1 | 2 | 5 | 1 | II |
| *Ranunculus flammula* | | | | | | | 5 | 5 | 5 | | | | | | | I |
| *Ranunculus repens* | 1 | 1 | 1 | | | | 5 | 5 | 5 | 1 | 1 | 1 | 3 | 5 | 1 | I |
| *Rubus fruticosus* | 4 | 10 | 1 | 4.1 | 5 | 1 | 3. 7 | 5 | 1 | 3.2 | 5 | 1 | 4. 6 | 5 | 1 | V |
| *Rubus idaeus* | 3.4 | 5 | 1 | 2.3 | 5 | 1 | 5 | 5 | 5 | 5 | 5 | 5 | 4.4 | 5 | 1 | II |

| Species | Microcommunity 1 | | | Microcommunity 2 | | | Microcommunity 3 | | | Microcommunity 4 | | | Microcommunity 5 | | | Constancy |
|---|---|---|---|---|---|---|---|---|---|---|---|---|---|---|---|---|
| | mean | max | min | mean | max | min | mean | max | min | mean | max | min | mean | max | min | |
| *Rumex acetosa* | | | | | | | 1 | 1 | 1 | | | | 1 | 1 | 1 | I |
| *Salix aurita* | 7 | 10 | 5 | 5.5 | 10 | 1 | 10 | 20 | 5 | 8.7 | 20 | 1 | 4.2 | 5 | 1 | V |
| *Scrophularia nodosa* | 2.3 | 5 | 1 | | | | 1 | 1 | 1 | | | | 1 | 1 | 1 | I |
| *Solidago canadensis* | | | | | | | | | | 1 | 1 | 1 | | | | I |
| *Sorbus aucuparia* | 1.7 | 5 | 1 | 1 | 1 | 1 | 2.3 | 5 | 1 | 2.6 | 10 | 1 | 1.6 | 5 | 1 | IV |
| *Stellaria glauca* | 3 | 5 | 1 | | | | | | | | | | 3 | 5 | 1 | I |
| *Stellaria longifolia* | 5 | 5 | 5 | | | | 5 | 5 | 5 | | | | 5 | 5 | 5 | I |
| *Stellaria media* | 1 | 1 | 1 | 1 | 1 | 1 | | | | 1 | 1 | 1 | 1 | 1 | 1 | I |
| *Tanacetum vulgare* | | | | | | | 1 | 1 | 1 | | | | | | | I |
| *Taraxacum officinale* | | | | 1 | 1 | 1 | 1 | 1 | 1 | 1 | 1 | 1 | 1 | 1 | 1 | I |
| *Torilis japonica* | 5 | 5 | 5 | | | | 1 | 1 | 1 | | | | | | | I |
| *Trientalis europaea* | 3.3 | 5 | 1 | 2.6 | 5 | 1 | 3. 7 | 5 | 1 | 2.9 | 5 | 1 | 3 | 5 | 1 | IV |
| *Trifolium repens* | | | | 1 | 1 | 1 | | | | | | | | | | I |
| *Tussilago farfara* | 3 | 5 | 1 | | | | | | | 1 | 1 | 1 | 5 | 5 | 5 | I |
| *Vaccinium myrtillus* | 1 | 1 | 1 | 3.8 | 10 | 1 | 3.4 | 5 | 1 | 6. 9 | 20 | 1 | 2 | 5 | 1 | III |
| *Veronica chamaedrys* | | | | 1 | 1 | 1 | | | | | | | 1 | 1 | 1 | I |
| *Veronica officinalis* | | | | | | | | | | | | | 1 | 1 | 1 | I |
| *Veronica scutellata* | 1 | 1 | 1 | | | | 1 | 1 | 1 | | | | | | | I |

## APPENDIX 2. MEAN, MAXIMUM AND MINIMUM VALUES OF COVER AND CONSTANCY OF HERBACEOUS SPECIES IN THE UNTHINNED FOREST

| Species | Microcommunity 1 | | | Microcommunity 2 | | | Microcommunity 3 | | | Microcommunity 4 | | | Microcommunity 5 | | | Microcommunity 6 | | | Constancy |
|---|---|---|---|---|---|---|---|---|---|---|---|---|---|---|---|---|---|---|---|
| | mean | max | min | mean | max | min | mean | max | min | mean | max | min | mean | max | mn | mean | max | min | |
| *Agrostis stolonifera* | 5 | 5 | 5 | 5 | 5 | 5 | | | | 7.5 | 10 | 5 | | | | | | | I |
| *Agrostis capillaris* | 9.2 | 20 | 5 | 7.5 | 20 | 5 | 15 | 20 | 5 | 28.6 | 60 | 10 | 10.3 | 20 | 5 | 15.8 | 50 | 5 | III |
| *Ajuga reptans* | 3 | 5 | 1 | | | | | | | 1 | 1 | 1 | | | | | | | I |
| *Alnus glutinosa* | | | | 5 | 5 | 5 | | | | | | | 1 | 1 | 1 | | | | I |
| *Anthoxanthum odoratum.* | | | | 5 | 5 | 5 | | | | | | | | | | | | | I |
| *Athyrium filix-femina* | 5 | 5 | 5 | | | | | | | | | | 3.3 | 5 | 1 | | | | I |
| *Betula pendula* | 4 | 10 | 1 | 1 | 1 | 1 | 3.7 | 5 | 1 | 1.7 | 5 | 1 | 3.6 | 5 | 1 | 3.8 | 10 | 1 | V |
| *Calamagrostis villosa* | 35 | 70 | 10 | 55 | 70 | 50 | 46.7 | 60 | 30 | 9.2 | 20 | 5 | 11.9 | 30 | 1 | 30 | 50 | 10 | V |
| *Calamagrostis epigejos* | 10 | 20 | 5 | 6.7 | 10 | 5 | 16.7 | 20 | 10 | 15.1 | 20 | 1 | 5.4 | 10 | 1 | 7.2 | 10 | 5 | V |
| *Calluna vulgaris* | | | | | | | 5 | 5 | 5 | | | | 6.2 | 10 | 1 | 4 | 5 | 1 | I |
| *Carex brizoides* | 15.1 | 80 | 30 | 16.7 | 20 | 10 | 7.5 | 10 | 5 | 13.3 | 20 | 10 | 8.5 | 20 | 5 | 3 | 5 | 1 | III |
| *Carex nigra* | | | | | | | 1 | 1 | 1 | 5 | 5 | 5 | | | | | | | I |
| *Carex pallescens* | | | | | | | | | | 5 | 5 | 5 | 1 | 1 | 1 | 1 | 1 | 1 | I |
| *Carpinus betulus* | 2 | 5 | 1 | 1 | 1 | 1 | | | | 3 | 5 | 1 | | | | 1 | 1 | 1 | I |
| *Chamaenerion angustifolium* | 2.3 | 5 | 1 | 1 | 1 | 1 | 1 | 1 | 1 | 2 | 5 | 1 | 1 | 1 | 1 | 2 | 5 | 1 | III |
| *Crataegus monogyna* | | | | | | | | | | 1 | 1 | 1 | | | | | | | I |

| Species | Microcommunity 1 | | | Microcommunity 2 | | | Microcommunity 3 | | | Microcommunity 4 | | | Microcommunity 5 | | | Microcommunity 6 | | | Constancy |
|---|---|---|---|---|---|---|---|---|---|---|---|---|---|---|---|---|---|---|---|
| | mean | max | min | mean | max | min | mean | max | min | mean | max | min | mean | max | mn | mean | max | min | |
| *Deschampsia caespitosa* | 3 | 5 | 1 | 5 | 5 | 5 | | | | 3.7 | 5 | 1 | 2.8 | 5 | 1 | 3 | 5 | 1 | II |
| *Deschampsia flexuosa* | 10 | 30 | 5 | 7.5 | 10 | 5 | 56.7 | 70 | 50 | 1 | 1 | 1 | 11.6 | 30 | 1 | 35 | 70 | 5 | IV |
| *Dryopteris carthusiana* | 3.3 | 5 | 1 | 2.6 | 5 | 1 | 5 | 5 | 5 | 3 | 5 | 1 | 2.3 | 5 | 1 | 2.3 | 5 | 1 | IV |
| *Dryopteris filix-mas* | | | | | | | | | | 1 | 1 | 1 | 5 | 5 | 5 | | | | I |
| *Epilobium ciliatum* | | | | 1 | 1 | 1 | | | | 3 | 5 | 1 | | | | | | | I |
| *Equisetum sylvaticum* | | | | 1 | 1 | 1 | | | | | | | 1 | 1 | 1 | | | | I |
| *Fagus sylvatica* | 3.7 | 5 | 1 | 3.7 | 5 | 1 | 3 | 5 | 1 | 4 | 10 | 1 | 1.6 | 5 | 1 | 2 | 5 | 1 | III |
| *Festuca gigantea* | | | | | | | | | | 10 | 10 | 10 | | | | | 0 | 0 | I |
| *Frangula alnus* | 1 | 1 | 1 | 1.8 | 5 | 1 | 3 | 5 | 1 | 1 | 1 | 1 | 1.4 | 5 | 1 | 1.7 | 5 | 1 | IV |
| *Galeopsis tetrahit* | 1 | 1 | 1 | | | | | | | 1 | 1 | 1 | 1 | 1 | 1 | | | | I |
| *Galium aparine* | | | | | | | 1 | 1 | 1 | 1 | 1 | 1 | | | | | | | I |
| *Hieracium* sp. | | | | 1 | 1 | 1 | | | | 1 | 1 | 1 | | | | 1 | 1 | 1 | I |
| *Hieracium pilosella* | | | | | | | 1 | 1 | 1 | 1 | 1 | 1 | | | | 1 | 1 | 1 | I |
| *Holcus lanatus* | | | | 3.7 | 5 | 1 | 3 | 5 | 1 | 12.5 | 30 | 5 | 2.3 | 5 | 1 | 5 | 5 | 5 | II |
| *Holcus mollis* | 5 | 5 | 5 | 5 | 5 | 5 | | | | 11.7 | 20 | 5 | 6.4 | 10 | 5 | 11.3 | 20 | 5 | II |
| *Hypericum perforatum* | 1 | 1 | 1 | | | | | | | 1 | 1 | 1 | | | | | | | III |
| *Juncus conglomeratus* | 1.8 | 5 | 1 | 3 | 5 | 1 | 3 | 5 | 1 | 5.2 | 10 | 1 | 1.3 | 5 | 1 | 2 | 5 | 1 | I |
| *Juncus effuses* | | | | 1 | 1 | 1 | | | | | | | 5 | 5 | 5 | 1 | 1 | 1 | I |

## Appendix 2. (Continued)

| Species | Microcommunity 1 | | | Microcommunity 2 | | | Microcommunity 3 | | | Microcommunity 4 | | | Microcommunity 5 | | | Microcommunity 6 | | | Constancy |
|---|---|---|---|---|---|---|---|---|---|---|---|---|---|---|---|---|---|---|---|
| | mean | max | min | mean | max | min | mean | max | min | mean | max | min | mean | max | mn | mean | max | min | |
| *Juncus tenuis* | 1 | 1 | 1 | 5 | 5 | 5 | | | | 1 | 1 | 1 | | | | | | | I |
| *Lapsana communis* | 1 | 1 | 1 | | | | | | | 1 | 1 | 1 | | | | | | | I |
| *Larix decidua* | | | | | | | | | | | | | 1 | 1 | 1 | 1 | 1 | 1 | I |
| *Leontodon autumnalis* | 1 | 1 | 1 | | | | 1 | 1 | 1 | | | | | | | | | | I |
| *Lotus uliginosus* | | | | | | | | | | 1 | 1 | 1 | | | | | | | I |
| *Luzula multiflora* | 1 | 1 | 1 | | | | | | | 1 | 1 | 1 | 2 | 5 | 1 | 5 | 5 | 5 | I |
| *Luzula pilosa* | 1 | 1 | 1 | 1 | 1 | 1 | 1 | 1 | 1 | 3 | 5 | 1 | 3 | 5 | 1 | 1 | 1 | 1 | I |
| *Lycopodium clavatum* | | | | | | | | | | 5 | 5 | 5 | 4 | 5 | 1 | 3 | 5 | 1 | I |
| *Lysimachia vulgaris* | 5 | 5 | 5 | 1 | 1 | 1 | 5 | 5 | 5 | 5 | 5 | 5 | 1 | 1 | 1 | 1 | 1 | 1 | II |
| *Maianthemum bifolium* | 2.3 | 5 | 1 | 1 | 1 | 1 | | | | 2.6 | 5 | 1 | 1 | 1 | 1 | 1 | 1 | 1 | II |
| *Melampyrum pretense* | | | | | | | | | | | | | 2.3 | 5 | 1 | | | | I |
| *Melica nutans* | | | | | | | | | | 5 | 5 | 5 | | | | | | | I |
| *Moehringia trinervia* | 1 | 1 | 1 | | | | | | | 1 | 1 | 1 | 3 | 5 | 1 | 1 | 1 | 1 | I |
| *Molinia caerulea* | 19 | 50 | 1 | 11.7 | 20 | 5 | 20 | 30 | 10 | 5 | 5 | 5 | 15.4 | 50 | 1 | 19.3 | 50 | 5 | III |
| *Mycelis muralis* | 3 | 5 | 1 | | | | | | | 3.7 | 5 | 1 | | | | | | | I |
| *Oxalis acetosella* | | | | 1 | 1 | 1 | 5 | 5 | 5 | 5 | 5 | 5 | 2.3 | 5 | 1 | | | | I |
| *Padus serotina* | | | | 1 | 1 | 1 | | | | 2.3 | 5 | 1 | 1 | 1 | 1 | 1 | 1 | 1 | I |
| *Picea abies* | 1 | 1 | 1 | | | | 1 | 1 | 1 | 1 | 1 | 1 | 1.8 | 5 | 1 | | | | I |
| *Pinus* | | | | 1 | 1 | 1 | 1 | 1 | 1 | 1 | 1 | 1 | 1 | 1 | 1 | 1 | 1 | 1 | II |

| Species | Microcommunity 1 | | | Microcommunity 2 | | | Microcommunity 3 | | | Microcommunity 4 | | | Microcommunity 5 | | | Microcommunity 6 | | | Constancy |
|---|---|---|---|---|---|---|---|---|---|---|---|---|---|---|---|---|---|---|---|
| | mean | max | min | mean | max | min | mean | max | min | mean | max | min | mean | max | mn | mean | max | min | |
| *sylvestris* | | | | | | | | | | | | | | | | | | | |
| *Poa pratensis* | | | | | | | | | | | | | | 0 | 0 | 5 | 5 | 5 | I |
| *Populus tremula* | 3.3 | 5 | 1 | 2.3 | 5 | 1 | 3 | 5 | 1 | 1 | 1 | 1 | 2.7 | 5 | 1 | 2 | 5 | 1 | IV |
| *Potentilla reptans* | | | | | | | | | | | | | 2.3 | 5 | 1 | 1 | 1 | 1 | I |
| *Prunus spinosa* | | | | | 0 | 0 | | | | | | | | | | 1 | 1 | 1 | I |
| *Pteridium aquilinum* | 12.5 | 20 | 5 | | 0 | 0 | 5 | 5 | 5 | 25 | 40 | 10 | 8.2 | 20 | 1 | 62.2 | 90 | 40 | II |
| *Quercus robur* | 2.3 | 5 | 1 | 1 | 1 | 1 | 1 | 1 | 1 | | | | 1.3 | 5 | 1 | 1 | 1 | 1 | III |
| *Quercus rubra* | | | | | | | | | | | | | 1 | 1 | 1 | | | | I |
| *Quercus petraea* | 1 | 1 | 1 | 1 | 1 | 1 | 3 | 5 | 1 | 1.7 | 5 | 1 | 1 | 1 | 1 | 1 | 1 | 1 | III |
| *Ranunculus repens* | | | | | | | | | | 5 | 5 | 5 | | | | | | | I |
| *Rubus fruticosus* | 7.8 | 20 | 5 | 2.5 | 5 | 1 | 2.3 | 5 | 1 | 16.7 | 50 | 5 | 3.6 | 10 | 1 | 3.7 | 5 | 1 | V |
| *Rubus idaeus* | 5 | 5 | 5 | | | | | | | 5 | 5 | 5 | 1.7 | 5 | 1 | 1 | 1 | 1 | II |
| *Rumex acetosa* | | | | 1 | 1 | 1 | | | | | | | | | | | | | I |
| *Rumex acetosella* | | | | | | | | | | 1 | 1 | 1 | | | | | | | I |
| *Salix aurita* | 5 | 5 | 5 | 2.6 | 5 | 1 | 2.3 | 5 | 1 | 3 | 5 | 1 | 2.2 | 5 | 1 | 2.5 | 5 | 1 | IV |
| *Salix caprea* | | | | 1 | 1 | 1 | | | | 5 | 5 | 5 | 1 | 1 | 1 | | | | I |
| *Scrophularia nodosa* | | | | | | | | | | | | | | | | 5 | 5 | 5 | I |
| *Sorbus aucuparia* | | | | | | | | | | | | | 1 | 1 | 1 | | | | I |
| *Stellaria glauca* | | | | | | | | | | 1 | 1 | 1 | | | | | | | I |
| *Stellaria longifolia* | | | | | | | | | | | | | 1 | 1 | 1 | | | | I |

## Appendix 2. (Continued)

| Species | Microcommunity 1 | | | Microcommunity 2 | | | Microcommunity 3 | | | Microcommunity 4 | | | Microcommunity 5 | | | Microcommunity 6 | | | Constancy |
|---|---|---|---|---|---|---|---|---|---|---|---|---|---|---|---|---|---|---|---|
| | mean | max | min | mean | max | min | mean | max | min | mean | max | min | mean | max | mn | mean | max | min | |
| *Stellaria media* | 5 | 5 | 5 | | | | | | | 7.5 | 10 | 5 | 1 | 1 | 1 | 1 | 1 | 1 | I |
| *Taraxacum officinale* | 1 | 1 | 1 | | | | 1 | 1 | 1 | | | | 1 | 1 | 1 | | | | I |
| *Trientalis europaea* | 3.2 | 5 | 1 | 3 | 5 | 1 | 2.3 | 5 | 1 | 3 | 5 | 1 | 1.7 | 5 | 1 | 4 | 5 | 1 | V |
| *Trifolium repens* | | | | | | | | | | | | | | | | 1 | 1 | 1 | I |
| *Vaccinium myrtillus* | 5 | 5 | 5 | 1 | 1 | 1 | 5 | 5 | 5 | | | | 10.2 | 50 | 1 | 4 | 5 | 1 | III |
| *Viola mirabilis* | 1 | 1 | 1 | | | | | | | 5 | 5 | 5 | | | | | | | I |

## The Phytocenotic Diversity of teppe Vegetation in the Systems of Dominant and Floristic Classifications

the important directions in formation of the basic elements of phytobiota s to develop a classification of vegetation of the observed area. The two main f vegetation classification became widespread among the phytocenologists: tocenotic (the dominant) and ecologo-floristic. Both approaches can be used for studies because each of them reveals the essence of plant communities from es. However, as stressed by J.R. Sheljag-Sosonko (Sheljag-Sosonko, 2007), this assification, which displays the coenotic role of species, is a true vegetation . It should therefore be the basis for vegetation monitoring (Ostapko, Prikhodko, ously this approach was made in the classification of feather-grass steppes of the dratjuk, Chuprina, 1992).

assification scheme developed on the ecologo-phytocenotic base the vegetation s of formations, formations and associations are singled out. Formations are y the main dominants, and associations are defined on the ratios of the dominant dominant. For multilayered vegetation associations were established through f dominant species of each layer. Phytocenotic diversity was investigated at the smallest units of vegetation differentiation, which typically have the greatest omogeneity and stability of composition and structure of plant communities, cide with the category of association in syntaxonomy of the dominant (Aleksandrova, 1969; Ipatov, 1999; Methodology..., 1991; Yurtsev, 1991). studies were performed by the routing method with the preparation of descriptions by the standard technique (Program..., 1974). The names of plants cording to modern nomenclature (Ostapko et al., 2010).

publication of prodromus of the natural vegetation of the South-East of Ukraine nant basis in 1995 (Ostapko, 1995) many new associations and formations were ich is partly reflected in the publications (Glukhov et al., 2010; Ostapko et al., , 2011, 2012; Ostapko, 2005, 2011; Ostapko, Polyakov, 2003; Ostapko, a, 2010; Ostapko et al., 2011; Kupryushyna et al., 2011; Regional, 2011; 999). Further study of phytocoenotic diversity has led to the construction of assification on the dominant basis (Prikhodko et al., 2012), the structure of which in the Table 1.

vegetation of the region is represented by 10 types, which allocated 30 classes of n total 2905 association of 540 vegetation formations were found in the region iple of dominance. This diversity of associations reflects multiple successional are formed due to anthropogenic transformations of vegetation.

ersity of extrazonal vegetation of broad-leaved forests, especially ravine oak proportion of the associations to the formation is 29.4, is widely represented. It's ith a complex storeyed forest ecosystem, which reflects the peculiarities of structure of ravine-frame systems.

In: Terrestrial Biomes
Editor: Marlon Nguyen
ISBN: 978-1-63484-625-7


*Chapter 4*

# Regularities and Features of Differentiation and Anthropogenic Transformation of Steppe Vegetation

***F. N. Lisetskii, V. K. Tokhtar, V. M. Ostapko, S. A. Prykhodko and T. V. Petrunova***
Belgorod State National Research University, Belgorod, Russia

## Abstract

The questions of phytobiota evolution and formation of steppes in the vast region of the south of East European plain are reviewed. The main stages, aspects and ways of flora formation under increased anthropogenic impacts on steppe ecosystems are identified. Isolation and differentiation of different flora types at the present stage of development are shown. Their structure is determined by individuality of natural and anthropogenic interaction. The classification scheme of vegetation on the ecological-phytocenotic basis is developed and the vegetation types, formation classes, formations and associations are pointed out, using the results of long-term steppe phytocenoses research. The phytocenotic diversity of steppe vegetation in the systems of dominant and floristic classifications is characterized. The floristic richness, phytocoenotic diversity, uniqueness, stenotopic features, endemism, relictness, area-marginality of syntaxons are defined in the article. Rare for the region plant communities were identified. Regularities and peculiarities of the geographic and edaphic distribution of different level syntaxons of the south of East European plain steppe vegetation were established. It is shown that flora gene pool that has been preserved over centuries in the barrows and derelict lands can be used for eco-efficiency assessment of programs for the zonal steppe vegetation reconstruction.

## Introduction

The steppes on Earth cover more than 6% of the land area and are one of the main biomes, accumulating energy resources in humus-rich chernozem soils, which ruthless exploitation leads to a widespread degradation. Steppe are characterized by high biotic diversity and domination of xerophilous herbaceous vegetation, with sod grasses in base. High floristic richness, complex spatial structure, polydominancy, pluristratal organisation, seasonal aspects change, summer dormancy are characteristic for steppe phytocoenoses.

Currently, the development of flora and vegetation of the steppe zone in modern conditions is influenced by the growing anthropogenic impact leading to a change in the natural distribution of elements in soils and water imbalance. In this regard, steppe ecosystems go through significant changes under the influence of edaphic, geochemical and andrological neofactors. The appearance of such artificial anthropogenic ecotopes, having no natural alternative, leads to the formation of specific plant communities and floras, adapted to the extreme environmental conditions.

The steppe is the least protected biome according to all indicators: the least coverage of protected areas, the least biome fraction in national protected natural areas, the least average area of protected areas, etc. For main subdivisions of the steppe biome the proportion of the area under protection is evaluated nationally – 3–10% and for the total biome and all levels of conservation is not exceeding 5% (Smelansky, Tishkov, 2012). If we evaluate the reduction in the area that take natural ecosystems, the steppes are the most affected by human activity biome in the temperate zone (Henwood, 1998; Groom et al., 2006).

The region of Azov-Donetsk under study covers a large part of the Donetsk upland, southern spurs of the Central Russian upland, the Eastern part of the Azov upland and the Black sea lowlands.

## Evolution and Formation of Steppe Phytobiota in the South of East European Plain

Differentiation of anthropogenically transformed flora types of the steppe zone occurs in time and in space. Their evolution proceed alongside with the development of the natural flora within the natural ecotopes. According to A.N. Krishtofovich (1945) the formation of typical steppe in the geological past originated either autochthonously, or with the participation of foreign elements and was connected primarily with the aridity of climatic conditions.

Flora of Ancient Midlands influenced significantly the steppe flora formation and especially the appearance of petrophyton later (Lavrenko, 1970). It is noted that currently petrophyton is heterogeneous and heterochronic formation (Burda, 1991). It has formed, likely due to ancient plant species of the Miocene-Pleistocene period, among which an important role was played by the autochthonous species of Ancient Midlands, though it may be partially due to anthropogenic origin of petrophyton. Unique floral set of psammophyton is recognized by many authors as quite ancient and highly endemic complex, formed in specific conditions of isolated sandy terraces due to climatic changes. The formation of halophilous



complex is associated with the littorals of the marine bas
spread across saline habitats inland.
Total synantropization of plant cover in the steppe
holocene neocomplexes of ruderal and weed-grown flora
of the Azov-Donetsk region was growing as a result of h
age. Local impact: gathering, farming, cattle breeding, the
cities has been replaced by more intensive forms: plowir
that contributed to the formation of pro-synanthropic floris
In the XIX century the industrial development has led
areas of vegetation, but also to development of enti
anthropogenic ecotopes in steppe communities (Tokhtar, F
of pro-synanthropic flora were the first anthropogenic grou
the influence of anthropogenic factor. The increase of inter
impact led to the formation of various stable types of anthr
Burda (1991) identified the following anthropotolerant flo
study region: flora of nature reserve fund territories, depl
self-regenerative flora, cultivated flora of semi-natural
agrophytocenoses and flora of technogenic ecotopes, havin
Thus, the formation of the modern flora in the ste
obviously, has common traits with the same floras in the
unification. Under these conditions, initially, as the resul
ammophilous, petrophilous, in particular, calcipetrophilou
coexist steppe, forest-steppe, ammophilous, petrophilou
floristic complexes, which became the "material" base f
vegetation cover in steppe zone.
At the ancient stage of flora development the main an
cutting, selective gathering and grazing, that lead to forma
disturbed by pasture and deforestation. This resulted
ecotopes: sorting of soils, salinization-desalinization, deve
processes. The next stage of anthropogenic evolution of veg
steppe zones has become a large-scale economic developm
the formation of agricultural landscape – a combination of a
Under these conditions, there was the intensification of
synanthropic complexes and steppe complexes disturbed b
modifications of ecosystems appeared (seeded forage lands
roads, artificial reservoirs, canals, irrigation ditches).
In this context we can observe global synantropiza
adventive species (Burda, Tokhtar, 1992; Wittig, Lenker,
2003), and vegetation halophitization in some cases. Anthr
to the expansion of artificial areas of ecotopes and spread of
of apophytes (Tokhtar et al., 2011; Tokhtar, Groshenko, 201



S

One of
monitoring
principles
ecologo-phy
monitoring
different sid
dominant cl
classificatio
2010). Previ
region (Kon
In the c
types, class
determined
and the sub
correlation
level of the
degree of h
which coin
classificatio
Phytocenoti
geobotanica
are given ac
After th
on the domi
revealed, wh
2007, 2008
Kupryushyn
Chupina, 1
vegetation c
is presented
Natural
formations.
on the prin
series, which
The div
forests, with
connected
geomorphic

**Table 1. Structure of the natural vegetation of the Azov-Donetsk region on the dominant basis**

| Vegetation type (number of associations) | Class of formations | The number | | A/F |
|---|---|---|---|---|
| | | formations (F) | associations (A) | |
| Steppe – Steppa (1163) | Typical steppe – Steppa genuina | 64 | 763 | 11,9 |
| | Shrub steppe – Steppa fruticosa | 14 | 135 | 9,6 |
| | Stony steppe – Steppa petrophyta | 25 | 123 | 4,9 |
| | Calciphyte steppes – Steppa calcephyta | 8 | 23 | 2,9 |
| | Meadow steppe – Steppa pratensis | 33 | 118 | 3,6 |
| | Desert steppe – Steppa deserta | 6 | 13 | 2,2 |
| | Clay steppe – Steppa argillosa | 4 | 10 | 2,5 |
| | Channery-psammophyte steppe – Steppa detritica | 5 | 11 | 2,2 |
| Petrophyte – Petro-phyta (144) | Cretophyte – Cretophyta | 21 | 65 | 3,1 |
| | Calciphyte – Calcephyta | 32 | 130 | 4,1 |
| | Granitophyte – Granitophyta | 19 | 56 | 2,9 |
| Tomillare (137) | Tomillares – Tomillares | 13 | 138 | 10,6 |
| Psammo-phytic – Psammo-phyta (162) | Riverine sands – Psammophyta subriparia | 17 | 58 | 3,4 |
| | Seaside sands – Psammophyta submarina | 12 | 22 | 1,8 |
| | Terrace sands– Psammophyta supraterrasae | 22 | 82 | 3,7 |
| | Coniferous forest– Silvae acicularea | 1 | 8 | 8 |
| Forest – Silvae (379) | Broad-leaved coniferous forests – Silvae aciculari-latifoliosae | 1 | 2 | 2 |
| | Small-leaved coniferous forests – Silvae aciculari-parvifoliosae | 3 | 16 | 5,3 |
| | Broad-leaved summer-green forest – Silvae folioaestilignosa | 12 | 353 | 29,4 |
| | Genuine Meadows – Prata genuina | 43 | 229 | 5,3 |
| Meadow – Prata (354) | Steppe meadows – Prata substepposa | 16 | 86 | 5,4 |
| | Swampy meadows – Prata paludosa | 19 | 49 | 2,6 |
| | Saline meadows – Prata galophyta | 42 | 172 | 4,1 |
| Saline – Galophyta (212) | Genuine salt marshes – Eugalophyta | 13 | 40 | 3,1 |
| | Natural aquatic – Vegetalia aquatica | 22 | 39 | 1,8 |
| Aquatic – Hydrophyta (99) | Coastal aquatic – Vegetalia subaquatica | 28 | 61 | 2,2 |
| | Eutrophic swamps – Paludeseutrophicae | 27 | 51 | 1,9 |
| Swamp – Paludes (54) | Mesotrophic swamps – Paludes mezotrophicae | 2 | 3 | 1,5 |
| Shrub – Frutectosa (47) | Mesophytic shrubs – Frutectosa mesophyta | 4 | 7 | 1,8 |
| | Xerophytic shrubs –Frutectosa xerophyta | 12 | 42 | 3,5 |

The intrazonal vegetation of saline soils and hydrophytic ecosystems, that occupy small area in the region, are represented less diverse.

Intrazonal meadow vegetation is typical for the river valleys and ravine-frame systems flowing into them, as well as for the coastal line of the Azov sea, and it is represented by 78 formations and 354 associations, many of which are certain stages in the succession series, often caused by the anthropogenic influence.

But the most diverse at the level of associations is zonal steppe vegetation, that is proved by the high ratio of associations to formations, that is 11.9. The richest in associations formations of the typical steppe: *Festuceta valesiacae* (114 associations), *Elytrigieta repentis* (54), Stipeta *capillatae* (51), *Stipeta lessingianae* (37), *Poeta angustifoliae* (35), *Bromopsieta ripariae* (34), *Stipeta grafianae* (33), *Stipeta ucrainicae* (27), *Stipeta tirsae* (26), *Elytrigieta trichophorae* (22), *Bromopsieta inermis* (18), *Stipeta dazyphyllae* (18).

Considerable diversity is characteristic not only for the typical steppe, but also for its climatic and edaphic variants. In particular, many associations of meadow steppe are formed in the highest part of the Donetsk ridge and along the slopes of northern exposition (*Elytrigieta repentis* – 10, *Poeta angustifoliae* – 10, *Galatelleta dracunculis* – 10, *Filipenduleta vulgaris* – 9, *Stipeta tirsae* – 8).

The high diversity of shrub steppes is also significant (*Caraganeta fruticis* – 41, *Amygdaleta nanae* – 32 associations, *Caraganeta scythicae* – 9 (Figure 1a), *Spiraeeta hypericifoliae* – 8, *Calophaceta wolgaricae* – 7) and petrophyte steppes *(Stipeta capillatae* – 13, *Crinitarieta villosae* – 13, *Jurineeta brachycephalae* – 11, *Stipeta lessingianae* – 9, *Lineta czerniaëvii* – 9, *Salvieta nutantis* – 9 (Figure 1b), *Stipeta graniticolae* – 6, they formed due to natural vegetation succession of rock outcrops towards the climax stage.

The vegetation of the rock outcrops is varied, with many of the plant communities of stenotopic character formed here, endemic (*Stipeta graniticolae* (Figure 1c), *Euphorbieta cretophilae, Onosmateta tanaiticae, Artemisieta nutantis, Helianthemeta cretophilae, Thymeta pseudogranitici,* etc.) and relic *(Calophaceta wolgaricae, Artemisieta hololeucae, Thymeta kondratjukii, Hedysareta cretacei, Erodieta beketowii,* etc.).

Petrophyton can be easily divided into cretophyte, calciphyte and granitophyte group according to the floristic composition and structure of the formations. These florocomplexes formed during florogenesis as edaphic variants of steppophyton for a long period of time with simultaneous allopatric and parapatric speciation in some genera (*Thymus, Scrophularia, Asperula, Jurinea, Linum, Elytrigia*, etc.).

Cretophyta are represented by many endemic stenotopic associations, such as *Anthericetum (ramosi) helianthemosum (cretophili), Bromopsietum (ripariae) schivereckiosum (mutabilis), Caricetum (humilis) elytrigiosum (stipifoliae), Centaureetum (ruthenicae) purum, Crinitarietum (villosae) koeleriosum (talievii), Diplotaxietum (cretaceae) pimpinellosum (titanophilae), Hedysaretum (cretacei) artemisiosum (hololeucae), Hedysaretum (cretacei) festucosum (cretaceae), Matthioletum (fragrantis) purum, Pimpinelletum (titanophilae) artemisiosum (hololeucae), Pimpinelletum (titanophilae) euphorbiosum (cretophilae), Scrophularietum (cretaceae) artemisiosum (hololeucae), Scrophularietum (cretaceae) thymosum (cretacei), Stipetum (joannis) anthericosum (ramosi), Stipetum (joannis) elytrigiosum (cretaceae), Thymetum (cretacei) helianthemosum (cretophili)* (Figure 1d) etc., widespread in Seversky Donets basin, mostly in its left-bank tributaries.

Calciphyte complex Calcephyta is richer, it's connected with the outcrops of limestones, sandstones and shales of different genetic types mainly on the Donetsk upland and a little in the Azov area. The specific communities are: formation of *Thymeta calcarei* (18 associations), associations of *Artemisietum (marshalliani) pimpinellosum (titanophilae), Botriochloetum (ischaemi) anthemidosum (subtinctoriae), Caricetum (supinae) linosum (czerniaëvii), Cephalarietum (uralensis) thymosum (calcarei), Crinitarietum (villosae) ephedrosum (distachiae), Festucetum (valesiacae) achilleosum (leptophyllae), Pimpinelletum (titanophilae) silenosum (supinae), Pimpinelletum (titanophilae) thymosum (calcarei), Rosetum (chrshanovskii) melicosum (transsylvanicae), Rosetum (subpygmaeae) hylotelephiosum (decumbentis), Silenetum (supinae) atraphaxiosum (frutescentis), Stipetum (capillatae) thymosum (calcarei), Thymetum (dimorphi) centaureosum (carbonatae),* etc.

Granitophyta focuses on the Azov upland and is represented by a complex of endemic communities with the participation of a number of relict species. The most varied formations in its composition are *Thymeta granitici* (15 associations), *Erodieta beketowii* (8), *Thymeta pseudogranitici* (6), *Thymeta dimorphi* (6). The unique associations are *Achilleetum (glaberrimi) purum, Achilleetum (leptophyllae) asperulosum (graniticolae), Aurinietum (saxatilis) festucosum (valesiacae), Festucetum (valesiacae) jurineosum (graniticae), Scrophularietum (donetzicae) melicosum (transsylvanicae), Stipetum (graniticolae) thymosum (granitici*).

It should be emphasised that tomillares, common on the outcrops of chalk and limestone, are characterized by a small number of formations, presented by a very wide variety of associations (associations to formations ratio is 10.6) and is caused by both natural successions and pastoral pressure on these communities. In particular, these are *Thymeta cretacei* (32 associations), *Artemisieta tanaiticae* (18), *Hyssopeta cretacei* (13), *Jurineeta brachycephalae* (12), *Onosmateta tanaitici* (11), *Artemisieta hololeucae* (10), *Genisteta scythicae* (8), *Helianthemeta cretophili* (6).

Vegetation of the open sands is quite varied, but is poorer compared with other types. At the same time it is characterized by high specificity and endemic species are among the dominants. On the terraces of river valleys the sandy grasslands and pioneer vegetation of open sands are developed. The richest characteristic formations are: *Festuceta beckeri* (14), *Agropyreta lavrenkoani* (6), *Artemisieta tscherniaevianae* (6), *Cariceta colchicae* (6), *Koelerieta sabuletori* (5). In the coastal zone of the Azov sea the characteristic community formations of *Agropyreta lavrenkoani, Cakileta maritimae, Cariceta colchicae, Crambeeta ponticae, Cynancheta acuti, Ephedreta distachyae, Eringieta maritimi, Euphorbieta seguieranae, Festuceta beckeri, Glycyrrhizeta glabrae, Leymeta sabulosi* (Figure 1e) are widespread on the sandy, shell substrate. Psammophyte steppes at the outputs of paleogenic sands, occasionally seen in the region are of special attention. They are quite diverse, represented by such formations as *Festuceta beckeri* (18), *Stipeta borysthenicae* (13), *Koelerieta sabuletori* (9), *Cariceta colchicae* (5), etc.

a

b

c

Figure 1. (Continued).

d

e

f

Figure 1. Typical (b, d, e) and rare (a, c, f) plant communities occured in steppe zone of the South in the Middle Russian Upland: a) *Caraganeta scythicae,* b) *Salvieta nutantis,* c) *Stipeta graniticolae,* d) *Thymetum (cretacei) helianthemosum (cretophili),* e) *Leymeta sabulosi,* f) *Centaureeta ruthenicae.*

## Syntaxons of Xerothermic Steppe Petrophyte and Psammophyte Vegetation of the Region, Selected by the Method of Braun-Blanquet

At the present stage of phytobiota development the anthropogenic influence in the region is extending, the structure of ecosystems is changing, invasive species are brought (Burda, Tokhtar, 1992). But in general, the vegetation is characterized by a large diversity of plant communities, including endemic, as their dominants are local or regional endemics and subendemics, relict species, regionally rare species due to stenotropic conditions of extention of phytocenoses edificators, their border-areal location, as well as restrictions on their localization under the influence of anthropogenic factor. Rare fraction of phytocoenofund in the region accounts for about 15% of associations and 23% of formations of the dominant vegetation classification (Prikhodko et al., 2012).

Rare fraction of steppe and petrophyte vegetation is represented by syntaxons included in the Green book of Ukraine (Green ..., 2009), namely the following formation: *Genisteta scythicae* (8 associations), *Calophaceta wolgaricae* (7), *Caraganeta scythicae* (19), *Stipeta braunerii* (3), *Stipeta capillatae* (65), *Stipeta tirsae* (26), *Stipeta graniticolae* (6), *Stipeta borysthenicae* (13), *Stipeta zalesskii* (15), *Stipeta lessingianae* (44), *Stipeta grafianae* (35), *Stipeta joannis* (14), *Stipeta dasyphyllae* (19), *Stipeta ucrainicae* (28), *Amygdaleta nanae* (32), *Cariceta humilis* (5), *Elytrigieta stipifoliae* (13), *Glycyrrhiseta glabrae* (10), *Hyssopeta cretacei* (13), *Erodieta beketowii* (9), *Artemisieta hololeucae* (9), *Helianthemeta cretophili* (6).

Rare fraction of xerothermic vegetation also comprises regional rare plant communities, dominants and codominants of which are species subjected to special protection, and also extremely rare associations of stenotopic habitats. They include the following formations: *Achilleeta glaberrimi* (1 association), *Anemoneta sylvestris* (2), *Anthericeta ramosi* (2), *Artemisieta nutantis* (3), *Artemisieta tanaiticae* (18), *Astragaleta albicaulis* (2), *Aurinieta saxatilis* (2), *Cariceta pediformis* (3), *Centaureeta ruthenicae* (1) (Figure 1f), *Convolvuleta lineati* (2), *Cotoneastereta melanocarpi* (4), *Crambeeta ponticae* (1), *Diplotaxieta cretaceae* (2), *Ephedreta distachyae* (2), *Eringieta maritimi* (2), *Euphorbieta cretophilae* (5), *Festuceta cretaceae* (5), *Hedysareta grandiflori* (3), *Inuleta hirtae* (1), *Krascheninnikovieta ceratoidis* (2), *Matthioleta fragrantis* (1), *Onosmateta tanaitici* (15), *Paeonieta tenuifoliae* (8), *Roseta chrshanovskii* (2), *Roseta subpygmaeae* (5), *Scrophularieta cretaceae* (6), *Scrophularieta donetzicae* (1), *Scutellarieta creticolae* (2), *Sileneta cretaceae* (1), *Tamariceta gracilis* (7), *Teucrieta chamaedryos* (1), *Thymeta didukhii* (1), *Thymeta kondratjukii* (5), *Thymeta pseudogranitici* (6).

The general scheme of xerothermic steppe, petrophyte and psammophyte vegetation in the region according to the method of Braun-Blanquet (Solomakha, 2008; Tyschenko, 2006) includes the following syntaxons:

***Classis.* Festuco-Brometea Br.-Bl. et R.Tx. in Br.-Bl. 1949.**
***Ordo.* Festucetalia valesiacae Br.-Bl. et R.Tx. 1943.**
***Alliancia.* Festucion valesiacae Klika 1931.**
***Suballiancia.* Festucenion valesiacae Kolbek in Voravec et al. 1983.**
*Associatio.* Salvio nemorosae-Festucetum valesiacae Korotchenko et Didukh 1997.

*Associatio.* Festucetum rupicolae Soo 1940.

*Associatio.* Festuco valesiacae-Stipetum capillatae Sill. 1937.

*Associatio.* Plantagini stepposae-Stipetum pulcherrimae V.Solomakha 1995.

*Associatio.* Stipo ucrainicae-Agropyretum pectinatae Tyschenko 1996.

*Associatio.* Salvio nemorosae-Elytrigietum intermediae Tyschenko 1996.

*Associatio.* Euphorbio segieranae-Koelerietum cristatae Smetana, Derpoluk, Krasova 1997.

*Associatio.* Medicago romanicae-Crinitarietum villiosae Smetana, Derpoluk, Krasova 1997.

*Associatio.* Festucetum valesiacae Solodkova et al. 1986.

*Associatio.* Festuco valesiacae-Caraganetum frutici Smetana, Derpoluk, Krasova 1997.

*Associatio.* Festuco valesiacae-Koelerietum cristatae Smetana, Derpoluk, Krasova 1997.

***Suballiancia.* Achilleo setaceae-Poenion angustifoliae Tkachenko, Movchan et V.Solomakha 1987.**

*Associatio.* Medicago romanicae-Poetum angustifoliae Tkachenko, Movchan et V.Solomakha 1987.

*Associatio.* Achilleo setaceae-Poetum angustifoliae Marjuschkina et V.Solomakha 1986.

*Associatio.* Elytrigio trichophorae-Poetum angustifoliae (Kost. et al. 1984) V.Solomacha 1995.

*Associatio.* Verbasco lychnitis-Koelerietum cristatae Smetana, Derpoluk, Krasova 1997.

***Suballiancia.* Coronillo variae-Poenion angustifoliae Smetana, Derpoluk, Krasova 1997.**

*Associatio.* Coronillo variae-Poetum angustifoliae Smetana, Derpoluk, Krasova 1997.

***Alliancia.* Cirsio-Brachypodion pinnati Hadac et Klika 1994 em Krausch 1961.**

*Associatio.* Thymo marschalliani-Caricetum praecocis Korotchenko et Didukh 1997.

***Alliancia.* Fragario viridis-Trifolion montani Korotchenko et Didukh 1997.**

*Associatio.* Medicago-Festucetum valesiacae Wagner 1940.

*Associatio.* Betonico officinalis-Trifolietum montani Popova in Popova et al. 1986.

*Associatio.* Salvio pratensis-Poetum angustifoliae Korotchenko et Didukh 1997.

*Associatio.* Veronico austriacae-Chamaecythisetum austriaci Korotchenko et Didukh 1997.

*Associatio.* Agrimonio eupatoriae-Galietum ruthenici Smetana, Derpoluk, Krasova 1997.

*Associatio.* Bromopsio ripariae-Plantagetum lanceolatae Smetana, Derpoluk, Krasova 1997.

***Alliancia.* Astragalo-Stipion Knapp 1944.**

*Associatio.* Stipetum pennatae R. Jovanovic 1956.

*Associatio.* Astragalo austriaci-Salvietum nutantis Korotchenko et Didukh 1997.

*Associatio.* Thymo marschalliani-Crinitarietum villosae Korotchenko et Didukh 1997.

*Associatio.* Stipetum lessingianae Soo 1948.

*Associatio.* Vinco herbaceae-Caraganetum fruticis Korotchenko et Didukh 1997.

*Associatio.* Eryngio campestri-Achilletum nobilis Smetana, Derpoluk, Krasova 1997.

*Associatio.* Marrubio praecoci-Euphorbietum stepposae Smetana, Derpoluk, Krasova 1997.

***Alliancia.* Artemisio-Kochion Soo 1969.**

*Associatio.* Agropyro pectinato-Kochietum prostratae Zolyomi 1958 corr. Soo 1959.

***Alliancia.* Artemisio marschalliani-Elytrigion intermediae Korotchenko et Didukh 1997.**

*Associatio.* Astragalo dasyanthi-Elytrigietum intermediae Korotchenko et Didukh 1997.

***Alliancia.* Chamaecythision ruthenici Smetana, Derpoluk, Krasova 1997.**

*Associatio.* Plantagini stepposae-Chamaecytisetum ruthenici Smetana, Derpoluk, Krasova 1997.

*Associatio.* Potentillo argenteae-Thymetum dymorphi Smetana, Derpoluk, Krasova - 1997.

*Associatio.* Crinitario villosae-Chamaecytisetum ruthenici Smetana, Derpoluk, Krasova 1997.

***Classis.* Helianthemo-Thymetea Romaschenko, Didukh et V.Sl. 1996.**

***Ordo.* Thymo cretacei-Hyssopetalia cretacei Didukh 1989.**

***Alliancia.* Artemisio hololeucae-Hyssopion cretacei Romaschenko, Didukh et V.Sl. 1996.**

*Associatio.* Artemisio nutantis-Plantaginetum salsae Didukh 1989.

*Associatio.* Artemisio hololeucae-Polygaletum cretaceae Didukh 1989.

*Associatio.* Onosmo tanaiticae-Androsacietum kozo-poljanskii Romaschenko, Didukh et V.Sl. 1996.

*Associatio.* Scrophulario cretacei-Helianthemetum cretacei Romaschenko, Didukh et V.Sl. 1996.

***Alliancia.* Euphorbio cretophilae-Thymion cretacei Didukh 1989.**

*Associatio.* Jurineo brachicephalae-Helianthemetum cretophilae Romaschenko, Didukh et V.Sl. 1996.

*Associatio.* Euphorbio cretophilae-Jurinetum brachicephalae Didukh 1989.

***Alliancia.* Centaureo carbonatae-Koelerion talievii Romaschenko, Didukh et V.Sl. 1996.**

*Associatio.* Jurineo brachicephalae-Koelerietum talievii Romaschenko, Didukh et V.Sl. 1996.

*Associatio.* Gypsophilo oligospermae-Campanuletum sibiricae Romaschenko, Didukh et V.Sl. 1996.

*Associatio.* Bupleuro falcatae-Stipetum capillatae Romaschenko, Didukh et V.Sl. 1996.

*Associatio.* Androsacio kozo-poljanskii-Caricetum humilis Korotchenko et Didukh 1997.

***Classis.* Glycyrrhizetea glabrae V. Golub et Mirkin in V. Golub 1995.**

***Ordo.* Glycyrrhizetalia glabrae V. Golub et Mirkin in V. Golub 1995.**

***Alliancia.* Glycyrrhizion glabrae V. Golub et Mirkin in V. Golub 1995.**

*Associatio.* Glycyrrhizetum glabrae Tyschenko 1998.

***Classis.* Ammophiletea Br.-Bl. et R.Tx. 1943.**

***Ordo.* Elymetalia gigantei Vicherek 1971.**

***Alliancia.* Elymion gigantei Morariu 1957.**

*Associatio.* Elymio-Astrodaucetum littoralis Korzh., Volkova et Klukin 1984.

*Associatio.* Tournefortietum sibiricae Popescu et Sanda 1975.

*Associatio.* Elymetum gigantei Morariu 1957.

*Associatio.* Salsoletum sodae Slavnic 1939.

*Associatio.* Artemisietum arenariae Popescu et Sanda 1975.

*Associatio.* Crambo pontici-Leymetum sabulosi Tyschenko 1998.

*Associatio.* Agrostio maeoticae-Gypsophillietum perfoliatae Umanets O.Yu., Voityuk B.Yu., Solomakha I.V. 2001.

*Associatio.* Secalio-Seselietum tenderiense Umanets O.Yu., Voityuk B.Yu., Solomakha I.V. 2001.

***Classis.* Festuco-Limonietea Karpov et Mirk. 1986.**

***Ordo.* Festuco-Limonietalia Mirk. in V.Golub et V.Sl. 1988.**

***Alliancia.* Festuco-Limonion gmelini Mirk. in V.Golub et V.Sl. 1991.**

*Associatio.* Salvio tesquicolae-Koelerietum cristatae Saveljeva et V.Golub 1991.

***Classis.* Agropyretea repentis Oberd., Th.Mull. et Gors in Oberd. et al. 1967.**

***Ordo.* Agropyretalia repentis Oberd., Th.Mull. et Gors in Oberd. et al. 1967.**

***Alliancia.* Convolvulo-Agropyrion repentis Gors 1966.**

*Associatio.* Agropyretum repentis Gors 1966.

*Associatio.* Anisantho-Artemisietum austriacae Kost. 1986.

*Associatio.* Calamagrostietum epigeios Kost. in V.Sl. et al. 1992.

***Classis.* Festucetea vaginatae Soo 1968 em Vicherek 1972.**

***Ordo.* Festucetalia vaginatae Soo 1968 em Vicherek 1972.**

***Alliancia.* Festucion beckeri Vicherek 1972.**

*Associatio.* Centaureo odessanae-Festucetum beckeri Vicherek 1972.

*Associatio.* Anisantho tectori-Medicagetum kotovii Tyschenko 1996.

*Associatio.* Anisantho tectori-Medicagetum kotovii syntrichietosum ruralis Tyschenko 2000.

*Associatio.* Centaureo odessanae-Caricetum colchicae Tyschenko 1999.

*Associatio.* Anisantho tectori-Helichrysetum arenarii Tyschenko 1999.

*Associatio.* Inulo sabuletori-Rumicetum acetoselliae Umanets, Solomakha 1999.

***Alliancia.* Verbascion pinnatifidii Korzh. et Kljukin 1990.**

*Associatio.* Astragalo borysthenici-Ephedretum Korzh. et Kljukin 1990.

*Associatio.* Bassio laniflorae-Bromion tectorum (Soo 1957) Borhidi 1996.

*Associatio.* Secali sylvestris-Brometum tectorum Hargitai 1940.

Thus, the ecological-floristic classification of the xerophytic vegetation of the region consists of 7 classes, 7 orders, 16 alliances, 3 suballiances, 67 associations. Among them the proper steppe type includes plant communities of 3 classes (*Festuco-Brometea, Agropyretea repentis, Glycyrrhizetea glabrae*), petrophyte vegetation includes 1 class (*Festuco-Brometea*), tomillares include 1 class (*Helianthemo-Thymetea*), psammophyte vegetation include 3 classes (*Ammophiletea, Festucetea vaginatae, Festuco-Limonietea*). This scheme demonstrates that method of Braun-Blanquet gives a very fragmentary study of the vegetation in the region. In particular, there is no data on the syntaxonomic structure of vegetation, formed at the outputs of granites, limestones, paleogene sands, carbon shale – ecotopes, which are characterized by specific floristic complexes containing paleo- and neoendemics types, vicarious and border areal elements.

In accordance with the ecological-floristic classification it is difficult to distinguish plant communities requiring special measures of protection, as even the associations are rather extensive units, including both common and rare plant communities in terms of the dominant approach. Special protection can be attributed to the class *Glycyrrhizetea glabrae*, alliance *Artemisio hololeucae-Hyssopion cretacei*, associations of *Plantagini stepposae-Stipetum pulcherrimae, Stipetum pennatae, Stipetum lessingianae, Astragalo borysthenici-Ephedretum,*

*Jurineo brachicephalae-Helianthemetum cretophilae, Androsacio kozo-poljanskii-Caricetum humilis, Crambo pontici-Leymetum sabulosi.*

## IDLE LANDS AND BARROWS AS POTENTIAL SOURCES OF STEPPE VEGETATION RESTORATION

The analysis of land management forms connected with plant substance alienation in zonal phytocenosis allows us to correct productivity changes caused by an anthropogenic factor in different historic and ecological periods (Lisetskii, Chernyavskikh, Degtyar,' 2011).

The reduction of farming lands has become a worldwide trend since the middle of last century. During the period of 1961-2003, there were left up to 223 million hectares of arable land, most of which (58.3 million hectares) were left in Russia. Till the end of XX century the process of arable land reduction in Russia was a part of the national economy crisis, caused by political and socioeconomic transformations that took place in Russia in 1990s. Since the beginning of the XXI century, according to the official statistics, in the European part of Russia the dynamics of the acreage reduction at first slowed down and then balanced at the level of 50 million hectares. The study of cultivated lands of the European part of Russia with the use of geoinformation technologies based on the analysis of Earth remote sensing data also confirm this trend (Schierhorn et al., 2013; Prishchepov et al., 2014; Kitov, Tsapkov, 2015).

The study of long-term idle lands is of special scientific interest. The territory on the Tarkhankut Peninsula in north-west Crimea was occupied by nomadic indigenous communities and reveal that the site shared the fate of the entire Chersonesean *chora*, meeting a violent end in the early part of the third century BC (Stolba and Andresen, 2015). Here and in other northern regions of the Black Sea area there were found post-antique idle lands that have been discovered due to the wide availability of highly detailed, multi-temporal, frequently updated, space imagery data, results of which made possible a review study of boundary systems in the area of the ancient statehood of the Northern Black Sea region (Lisetskii et al., 2013; Lisetskii, Rodionova, 2015; Lisetskii, Stolba, Marinina, 2015; Smekalova et al., 2015).

**Table 2. The structure of epiterranean layer phytomass in terms of virgin and idle lands (May) (Lisetskii et al., 2015)**

| Type of land[a] | H, cm | TPC, % | Number of plant species | Composition of herbage[b] | Mass of dry substance[c], $g/m^{-2}$ | | |
|---|---|---|---|---|---|---|---|
| | | | | | *F* | *R* | *SC* |
| VL | 36 | 75 | 17 | 1+2+14 | 159.16 | 254.52 | 182.00 |
| PIL | 32 | 95 | 12 | 4+2+6 | 85.56 | 181.68 | 112.04 |
| MIL | 29 | 70 | 12 | 2+1+9 | 118.88 | 117.56 | 61.00 |

Abbreviations: H – height of herbage; TPC – total projective cover.

[a]VL – virgin lands; PIL – post-antigue idle land; MIL – idle land in the modern era.

[b] The numbers specify respectively graminoids + legumes + herbs.

[c] F – green phytomass; R – plant debris; SC – litter.

**Table 3. Higher vegetation phytomass supply (g/m-2) in terms of idle lands (October)**

| Type of land | H, cm | epiterranean phytomass | | subterranean phytomass in layer, sm | | | |
|---|---|---|---|---|---|---|---|
| | | *F+R* | SC | 0-10 | | 0-20 | |
| | | | | green mass [b] | mortmass | green mass [b] | mortmass |
| | Ground 1 | | | | | | |
| MIL | 94 | 723.2 | 337.8 | 1273.7 | 1140.7 | 1556.3 | 1649.8 |
| PIL | 66 | 323.0 | 186.7 | 632.8 | 795.7 | 753.8 | 972.4 |
| MIL | 110 | 194.8 | 239.5 | 216.7(354.2) | 529.3 | 328.5(415.7) | 665.6 |
| | Ground 2 | | | | | | |
| PIL | 40 | 134.4 | 68.0 | 1349.9(136.6) | 1088.9 | 1508.2(136.6) | 1355.8 |
| PIL | 50 | 127.4 | 77.4 | 1956.2 | 959.2 | 2081.3 | 1196.9 |
| MIL | 43 | 88.6 | 187.9[a] | 498.5(83.5) | 786.6 | 709.1(83.5) | 954.0 |

Abbreviations. [a] Besides, ground litter contains sheep excrement weighing 50.2 g/m$^{-2}$.
[b] The additional weight of roots are noted within brackets (diameter > 0.6 mm).

Reconstructive successions of multi-temporal idle lands have the structure of the epiterranean layer phytomass, which is different from the indigenous communities (Tables 2-3).

The proportion of steppe grasses, which transfer from rootstock grasses to sod grasses stage of demutation with idle land aging, is the biggest in post-antique idle lands (*Stipa sp., Festuca valesiaca Gaudin, Bromopsis cappadocica (Boiss. and Balansa) Holub*) and new idle land (*Stipa lessingiana Trin. and Rupr., Koeleria cristata (L.) Pers.*), and in terms of virgin land one dominant species remains (*Stipa capillata L.*).

In spring (in May) when fescue, and not feather, as it will be later, becomes the main edificator of steppe (Table 1), the above-ground dry mass (F+R) of multiple-aged idle lands is 39% inferior to virgin soil. In autumn (October) in low grazing conditions on black earth soils, when the role of *Stipa capillata* becomes dominant, the weight (F+R) is by 55% (post-antique idle lands) and 73% (new idle lands) less than in virgin soils (Table 2).

The deposition of the aboveground mortmass (R+SC) depends on the magnitude of maximum green mass (F) and the speed of mortmass decomposition. The more active the production process is, a relative measure of which is F, and slower the speed of destruction, the more aboveground mortmass is accumulated. In these conditions, the maximum aboveground mortmass is noted in the terms of virgin land (436 g/m$^{-2}$), and in idle lands it decreases, and also new idle lands are inferior in this index to post-antique idle lands in 1.6 times.

The landscape of virgin steppe is inseparable from barrows, which were dominant, occupying the commanding heights. This remarkable feature of the steppe was noted by its first explorers. There is a great number of barrows in the steppes, but the rate of their loss is progressive. For example, in the Volgograd region there are more than 200,000 barrows (176 per 100 km$^2$), in Kalmykia there are 70 000 barrows (92 per 100 km$^2$), on the flat part of the Crimea (59 per 100 km$^2$). The number of barrows gets a sequence higher, if low barrows are found. Often they can be identified on lands with long-term cultivation and then magnetic investigations is used (Smekalova et al., 2005).

A barrow is a local geocomplex with dome-shaped top facies with xeromorphic vegetation, often anthropogenically disturbed, differently exposed slopes and a ditch with mesophytic vegetation. As a result of applicative evolution, the unique soils coordinated by topogradient in catena, are formed in the humus barrow fill. There still can be found the giant barrows of the steppe Scythia. For example, Oguz and Chertomlyk, volume of each reaches 8 thousand m$^3$. They are not only the unique monuments of archeology, but undoubtedly the most remarkable monuments of nature. The synthesis of complex natural science results of barrows research (Barczi, 2003; Sudnik-Wójcikowska, Moysiyenko, 2012 et al.) reveals the potential of barrows as non-reproducible natural ecosystem models for exploring a wide range of scientific problems.

High barrows, like the ones we studied, are not cultivated, they are surrounded by agricultural landscapes and represent island ecosystems, which have inherited particular flora features of previously undisturbed ecological background and have specific soil cover. It is a temporary stage of achieving soil climax with inherited properties of transplanted and mixed soils. Using the results of the preliminary surveys 106 barrows with a height of > 4 m, for which the typical vegetation was diagnosed by the presence of steppe flora and vegetation elements, especially turf grasses of the genera *Stipa, Festuca, Koeleria* (Sudnik-Wójcikowska, Moysiyenko, 2012), the most representative objects of study in transzonal view were identified. They are a well-preserved earth tombs with a height of 6-7.5 m, which are located in climatically different conditions of forest-steppe and steppe zones of the East European plain (Lavrenko, Karamysheva, Nikulina, 1991): in Cherkasy (F), Nikolaev (R) and Kherson (P, D) regions of Ukraine (Figure 2).

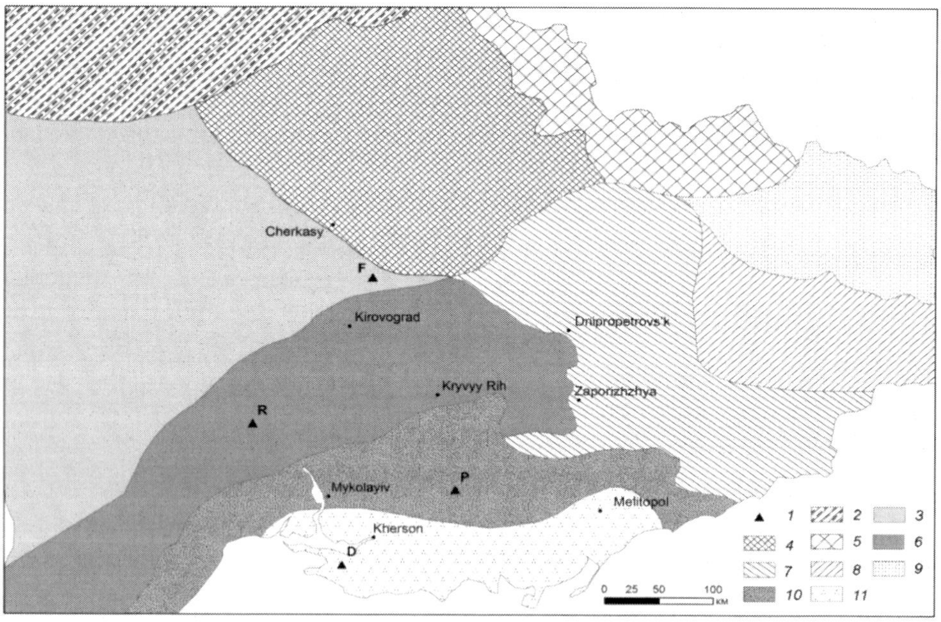


Figure 2. The location of the key objects of study – barrows in the forest-steppe and steppe in the scheme of physical-geographical zoning. *1* – barrows (F, R, P, D); *2* – Polesie province (coniferous-broadleaved zone). Forest-steppe zone: *3* – Dniester-Dnepr, *4* – Left-Bank Dnepr, *5* – Central Russian forest-steppe province. The steppe zone: *6* – Dniester-Dnepr, *7* – Left-Bank Dnepr-Azov, *8* – Donetsk, *9* – Donetsk-Don north steppe province (north steppe subzone). *10* – Black Sea middle steppe province (middle steppe subzone). *11* – Black Sea-Azov steppe province (dry steppe subzone).

For quite a long period of renaturation barrows form soil-vegetable cover, by topogradient corresponding to microlandscapes conditions of ecotopes: the heights, differently exposed slopes and foothills. Achieving a state of climax in plant community is determined by the ecosystem component with the largest characteristic time – edaphotope (Lisetskii, 1998). Known from the literature time estimations of full recovery of plant communities at idle lands overgrowing vary considerably from 50 to 200 years. It is therefore important to assess the degree of maturity of both plant communities and soils which recovered in different ways in separate ecotypes of a barrow since the construction of embankments.

The developed method of soil-genetic chronology (Goleusov, Lisetskii, 2008) is a new method of age determination of anthropogenic structures based on mathematical dependencies of irreversible genetic soil properties on time, determines the relevance of soil science for the attribution and protection of cultural heritage. In particular, this method allows to determine the last time humus material was added to the barrows (Lisetskii, 2012). However, in addition to absolute age of different barrow locations the soils and plant communities differ in relative age.

The analysis of the similarities and differences of individual edaphotopes by the combination of soil properties (agrochemical properties directly relevant to plant growth (8 indicators) and geochemical properties (18 indicators)) indicates some trends in the formation of specific consolidated groups on the southern slopes, which the tops of the barrows tend to. The northern slopes are of a great variety of properties. In separate barrow ecotopes due to the objectively existing differences in age ("maturity") of soils and vegetation the individual development course is implemented. Thus, the slopes of the barrows are formed by the maximum (in comparison with other barrow ecotypes) representation of steppe vegetation classes in the phytosociological spectrum (Lisetskii et al., 2014). Therefore, the slope locations of barrows contain the most valuable genetic and coenotic flora fund, which is useful for ecological restoration of the zonal steppe vegetation in conditions of flat interfluves.

## CONCLUSION

Thus, xerothermic steppe vegetation in the Azov-Donetsk region is very diverse due to the diversity of ecotopes, long-term evolution and anthropogenic influence. The formation of the modern flora in the steppe zone occurred gradually and, obviously, has something in common with the same floras in the other regions, indicating their unification. Under these conditions, initially, in the evolution of vegetation, specific ammophilous, petrophilous, in particular, calcipetrophilous complexes formed. Currently there coexist steppe, forest-steppe, ammophilous, petrophilous, halophilous and hydrophilous floristic complexes, which became the "material" for the present anthropogenic evolution of vegetation cover in steppe zone. Its natural differentiation is associated with geomorphological and geological factors, which caused subclimax vegetation formation on undeveloped and washed away soils, underlain by sands and sandstone of various genesis, chalk, limestone and granites. This led to the formation of a specific floristic complexes and corresponding syntaxons of vegetation. Due to anthropogenic factors successional series of vegetation increasing the diversity of plant communities and complicating their classification have formed. In sozological regard the vegetation of the region is saturated with syntaxons that are subject to special protection.

In separate barrow ecotopes due to the objectively existing differences in the age of the soils and vegetation the individual development course is implemented. The southern slopes always form the most favourable conditions for the continuous combination and recombination of species in the community. Slope ecotope of the barrow is characterized by a maximum (in comparison with other barrow ecotypes) representation of steppe vegetation classes in the phytosociological spectrum. Therefore, the slope locations of barrows contain the most valuable coenotic of flora fund, which is useful for ecological restoration of the zonal steppe vegetation in conditions of flat interfluves.

## References

Alexandrova, VD. The classification of vegetation. Leningrad: *Nauka*, 1969. (in Russian).

Barczi, A. Data for the botanical and pedological surveys of the Hungarian kurgans (Great Hungarian Plain, Hortobagy). *Thaiszia–J. Bot.*, 2003 v 13, 113-126.

Burda, RI. Anthropogenic transformation of flora. Kiev: Naukova Dumka, 1991. (in Russian).

Burda, RI; Tokhtar, VK. Invasion, distribution and naturalization of plants along railroads of the Ukrainian south-east. *Ukr. Botan. Journ.*, 1992. v 49 no 5, 14-18.

Chuprina, TT. Sintaxonomic variety psammophytic steppes in the south-east of Ukraine. Introduction and acclimatization of plants, 1999 no 32, 116-121. (in Russian).

Glukhov, AZ; Ostapko, VM; Prikhodko, CA. Phytodiversity regional landscape park "Meotida." Landscapes, vegetation and fauna of the regional landscape park "Meotida." Donetsk: N*oulidzh* 2010, 15-78. (in Russian).

Goleusov, PV; Lisetskii, FN. Soil development in anthropogenically disturbed forest-steppe landscapes. *Eurasian Soil Science,* 2008 v 41 no 13, 1480-1486. DOI: 10.1134/ S1064229308130188.

Green Book of Ukraine. Didukh YP, editor. Kyiv: Alterpres, 2009. (in Ukrainian).

Groom, MJ; Meffe, GK; Carroll, CR. Principles of conservation biology (pp. 174-251). Sunderland: Sinauer Associates. 2006.

Henwood WD. An overview of protected areas in the temperate grasslands biome. *Parks,* 1998 v 8 no 3, 3-8.

Ipatov, VC. Phytocenology. SPb: Publishing House of St. Petersburg University Press, 1999. (in Russian).

Kitov, MV; Tsapkov, AN. Assessment of the area of fallow land in the Belgorod Region and other regions of European Russia for the period 1990-2013 years. *Belgorod State University Scientific Bulletin. Natural sciences*, 2015 no 15(212) issue 32, 163-171.

Kondratyuk, EN. Feather grass steppes of Donbass. Kiev: *Naukova Dumka,* 1992. (in Russian).

Krishtofovich AN. The main way of development of flora and vegetation in the Cenozoic (the presentation of the report at a meeting of the Standing Committee on the history of flora and vegetation of the USSR in 1945). *Soviet botany,* 1945 no 5, 47-48. (in Russian).

Kupryushyna, LV.; Ostapko, VM; Kolomiychuk VP. Phytocoenotic features *Caragana scythica (Kom.) Pojark. (Fabaceae Lindl.)* in Donetsk-Azov region. *Chernomorskyy Botanical Magazine*, 2011 v 7 no 3, 238-252. (in Ukrainian).

Lavrenko, EM. The provincial division of the Black sea-Kazakhstan subregion steppe region of Eurasia. *Botanical Magazine*, 1970, v 55 no 5, 609-625. (in Russian).

Lavrenko, EM; Karamysheva, ZV; Nikulina, RI. *Steppes of Eurasia*. Leningrad: *Nauka*; 1991. (in Russian).

Lisetskii, F; Stolba, V; Ergina, E; Rodionova, M; Terekhin, E. Post-agrogenic evolution of soils in ancient Greek land use areas in the Herakleian Peninsula, South-West Crimea. *The Holocene*, 2013 no 4, 504-514. DOI: 10.1177/0959683612463098.

Lisetskii, F; Stolba, VF; Marinina, O. Indicators of agricultural soil genesis under varying conditions of land use, Steppe Crimea. *Geoderma*, 2015 v 239-240, 304-316.

Lisetskii, FN. Autogenic succession of steppe vegetation in postantique landscapes. *Russ. J. Ecol.*, 1998 v 29 no 4, 217-219.

Lisetskii, FN. Soil reproduction in steppe ecosystems of different ages. *Contemporary problems of ecology*, 2012 v 5 no 6, 580-588. DOI: 10.1134/S1995425512060108.

Lisetskii, FN; Chernyavskikh, VI; Degtyar,' OV. Pastures in the zone of temperate climate: trends for development, dynamics, ecological fundamentals of rational use. Pastures: Dynamics, Economics and Management. Ed. by N.T. Procházka. *Nova Science Publishers,* Inc., 2011, 51-83.

Lisetskii, FN; Goleusov, PV; Moysiyenko, II; Sudnik-Wojcikowska, B. Microzonal distribution of soils and plants along the catenas of mound structures. *Contemporary Problems of Ecology*, 2014 v 7 no 3, 282-293.

Lisetskii, FN; Rodionova, ME. Transformation of dry-steppe soils under long-term agrogenic impacts in the area of ancient Olbia. *Eurasian Soil Science*, 2015 v 48 no 4, 347-358. DOI: 10.1134/S1064229315040055.

Ostapko, VM. Eydologic, population and cenotic population and phytosozology bases in the south-east of Ukraine. Donetsk Ltd. "Lebed,'" 2005. (in Russian).

Ostapko, VM. Phytobiota regards cenotic diversity in southeastern Ukraine and its protection. *Ukrainian Botanical Journal,* 1999 v 56 no 5, 536-543. (in Ukrainian).

Ostapko, VM. Prodromus natural vegetation of the south-east of Ukraine. *Donetsk,* 1995. (in Russian).

Ostapko, VM. Sozological assessment and protection of the natural vegetation of Azov region of Donetsk. Network key botanical areas in the Azov region. Kiev: *Alterpres* 2011, 27-30. (in Russian).

Ostapko, VM; Boyko, AV; Mosyakin, SL. Vascular plants of the south-east of Ukraine. Donetsk: *Noulidzh,* 2010. (in Russian).

Ostapko, VM; Glukhov, AZ; Blakbern, AA; et al., Donetsk regional ecological network areas: concept, program and location. Ostapko VM, editor. Donetsk: *Tehnopak,* 2008. (in Ukrainian).

Ostapko, VM; Kozub-Ptiza, VV; Ibatulina, YV; Hnatiuk, NY. Phytosozological assessment of the tract wide beam (Donetsk region). *Industrial botany*, 2011 no 11, 97-104.

Ostapko, VM; Kupryushina, LP; Mulenkova, EG. Sozological justification for the establishment of the botanical reserve "Gruzskolomovsky" (Donetsk region). *Industrial botany,* 2007 no 7, 85-90. (in Russian).

Ostapko, VM; Kupryushina, LV. Phytocoenotic variety of shrub steppes in the south-east of Ukraine and its sozological score. Industrial botany: status and prospects. *Donetsk,* 2010, 346-349. (in Russian).

Ostapko, VM; Mulenkova, EG; Hnatiuk, NU; Zybenko OV. Phytosozological study for the establishment of the regional landscape park "Skelevoy" (Donetsk region). *Industrial botany,* 2008 no 8, 62-69. (in Russian).

Ostapko, VM; Polyakov, AK. Phytosozological assessment of the regional landscape park "Zuevsky." *Industrial botany,* 2003 no 3, 44-51. (in Russian).

Ostapko, VM; Prikhodko, CA; Vereshchetin, IM. Phytocoenotic bases expand the territory of the regional landscape park "Donetsk ridge" and its zoning. Preservation, restoration and stabilization of steppe ecosystems. *Mariupol: Renata,* 2011, 96-102. (in Ukrainian).

Ostapko, VM; Sova, TV; Nazarenko, AS; Ibatulina, YV. Flora and vegetation Branch "Trehizbenskaya steppe" Luhansk Nature Reserve. *Industrial botany,* 2012 no 12, 67-74. (in Russian).

Prikhodko, CA; Ostapko, VM; Kupryushyna, LV. Syntaxonomical variety of vegetation in the South-East Ukraine aspect synfitosozology. *Industrial botany,* 2012 no 12, 53-60. (in Ukrainian).

Prishchepov, AV; Radeloff, VC; Dubinin, M; Alcantara, C. The effect of Landsat ETM/TM+ image acquisition dates on detection of agricultural land abandonment in Eastern Europe (accessed 15 November 2014). Available from: http://www.R-project.org.

Program and methods biogeocenological research. Dylis NV, editor. Moscow: *Nauka* (Science), 1974. (in Russian).

Red Book of Ukraine. Flora. Didukh YP, editor. Kiev: *Global consulting,* 2009. (in Ukrainian).

Schierhorn, F; Müller, D; Beringer, T. Post-Soviet cropland abandonment and carbon sequestration in European Russia, Ukraine, and Belarus. *Global Biogeochemical Cycles*, 2013 v 27, 1175-1185.

Shelyag-Sosonko, YP. Biodiversity: paradigm and definitions. *Ukrainian Botanical Journal,* 2007 v 64 no 6, 777-796. (in Ukrainian).

Shelyag-Sosonko, YP; Krisachenko, VS; Movchan, YI. Methodology geobotany. Kiev: *Naukova Dumka,* 1991. (in Russian).

Smekalova, T; Voss, O; Smekalov, S; Myts, V; Koltukhov, S. Magnetometric investigations of stone constructions within large ancient barrows of Denmark and Crimea. *Geoarchaeology*, 2005 v 20 no 5, 461-482.

Smekalova, TN; Bevan, BW; Chudin, AV; Garipov, AS. The Discovery of an Ancient Greek Vineyard. *Archaeological Prospection*, 2015. DOI: 10.1002/arp.1517.

Smelansky, IE; Tishkov, AA. The Steppe biome in Russia: Ecosystem services, conservation status, and actual challenges. Eurasian Steppes. Ecological problems and livelihoods in a changing world. 2012. *Springer* Netherlands, 45-101.

Solomaha, VA. Ukraine syntaxonomy vegetation. The third approach. Kiev: *Fytosociocentre,* 2008. (in Ukrainian).

Stolba, VF; Andresen, J. Unveiling the hinterland: a new type of Hellenistic rural settlement in Crimea. *Antiquity*, 2015 v 89(344), 345-360.

Sudnik-Wójcikowska, B; Moysiyenko, II. Kurhany na "Dzikich Polach" – dziedzictwo kultury i ostorja ukraińskiego stepu. Warszawa, 2012.

Tishchenko, OJ. Vegetation coastal spits northern coast of the Azov Sea. Kiev: *Fytosociocentre*, 2006. (in Ukrainian).

Tokhtar, VK; Groshenko, SA. Differentiation of the Climatic Niches of the Invasive Oenothera L. (Subsect. Oenothera, Onagraceae) Species in the Eastern Europe. *Advances in Environmental Biology*, 2014 v 8 no 10, 529-531.

Tokhtar, VK; Petin, AN. Evolution and differentiation of phytobiota under anthropogenous impact in steppe and steppe-forest zone. *Bulletin of the Russian Academy of Sciences. Srer. Geography*, 2012 no 6, 83-91.

Tokhtar, VK; Vinogradova, YK; Groshenko, SA. Microevolution and Invasiveness of Oenothera L. Species (Subsect. Oenothera, Onagraceae) in Europe. *Russian Journal of Biological Invasions*, 2011 v 2 no 4, 273-280.

Wittig, R; Lenker, K-H; Tokhtar, VK. Zur Sociologie von Arten der Gattung Oenothera L. im Rheintal von Arnheim (NL) bis MÜlhouse (F). *Tuexenia*, 1999 v 19, 447-467.

Wittig, R; Tokhtar, VK. Die Haufigkeit von Oenothera-Arten im westlichen Mitteleuropa. *Feddes Repertorium*, 2003 v 114 no 5-6, 372-379.

Yurtsev, BA. The study of biological diversity and comparative floristry. *Botanical Magazine,* 1991 v 76 no 3, 305-313. (in Russian).

## BIOGRAPHICAL SKETCH

**Name:** Fedor Lisetskii

**Affiliation:** The chairmen of the Environmental Management and Land Survey Chair, The director of the Federal-regional centre of aerospace and surface monitoring of the objects and natural resources,

Doctor of geography, Professor

**Date of Birth:** 03.08.1958

**Education:** Degrees: D.Sc. (Biogeography and Soil Geography), 1994; Ph.D. (Biogeography and Soil Geography), 1984. All three degrees were received from the Odessa State University (now - Odessa National University), Ukraine.

**Address:** Russian Federation, 308015
Belgorod, Pobeda Street, 85
Tel.: 7(4722) 301370
Fax: 7(4722) 301371
E-mail: liset@bsu.edu.ru

**Research and Professional Experience:**

| | |
|---|---|
| **Institution.** Belgorod State University<br>**Position**, The director of the Federal-regional centre of aerospace and surface monitoring of the objects and natural resources | 2006-present |
| **Institution.** Belgorod State University<br>**Position**, The chairmen of the Environmental Management and Land Survey | 1998-2010 |
| **Institution.** Belgorod State University | 1999- |

| | |
|---|---|
| **Position**, The chief of the department of sciences and research, vice-rector in science | 2002 |
| **Institution.** Belgorod State University | 1997- |
| **Position**, vice-director of the institute of natural-sciences problems (Belgorod | 1999 |
| **Institution.** Odessa State University | 1994-1995 |
| **Position**, vice-director of the South centre of agroecology | |
| **Institution.** Odessa State University | 1983-1984 |
| **Position**, researcher | |

**Research interests:** ecology, soil science, soil geography, geomorphology, geoarchaeology, pedoarchaeology, the study of ancient systems of land use new scientific methods (GIS, remote sensing).

**Professional Appointments:**

Author, co-author of over 400 publications including 7 books, 40 papers in peer-reviewed journals, and 15 of the objects of intellectual property (databases, computer programs).

**Honors:**

| | |
|---|---|
| diploma of the Russian ministry of Education | 2013 |
| Expert of the Russian ministry of Education | 2011-present |
| Expert of the Russian Foundation for Basic Research (RFBR) | 2011-present |
| Manager of the year in the nomination "Science" | 2008 |
| laureate of National ecological prize | 2007 |
| diploma holder of the National Ecological prize “EcoSpace” | 2005 |
| Expert of the Institute for Sustainable Communities | 2002 |
| Honorary worker of higher professional education of the Russian Federation | 2002 |
| diploma of the Russian ministry of Education | 2001 |
| full member (academic)of the International Science Academy of ecology and safety of the vital functions | 2000 |
| the corresponding member of Petrovskaya Academy of sciences and arts | 1999 |
| laureate of the South centre of scientific and technical activity contest, the investigations and social initiatives on the best scientific work in the sphere of ecology | 1990 |

**Publications Last 3 Years:**

***Journal Publications***

Lisetskii F, Stolba V, Ergina E, Rodionova M, Terekhin E. Post-Agrogenic Evolution of Soils in Ancient Greek Land Use Areas in the Herakleian Peninsula, South-West Crimea. The Holocene. 2013. №4. C. 504-514.

Lisetskii F. N., Goleusov P. V., Chepelev O. A. The Development of Chernozems on the Dniester–Prut Interfluve in the Holocene // Eurasian Soil Science, 2013, Vol. 46, No. 5, pp. 491–504. DOI: 10.1134/S1064 229313050086.

Lisetskii F.N., Pavlyuk Ya.V., Kirilenko Zh.A., Pichura V.I. Basin organization of nature management for solving hydroecological problems // Russian Meteorology and Hydrology. 2014. V. 39. N 8. P. 550-557.

New opportunities of geoplanning in the rural area with the implementing of geoinformational technologies and remote sensing / Fedor N. Lisetskii, Alla V. Zemlyakova, Edgar A. Terekhin, Anastasiya G. Naroznyaya, Yaroslava V. Pavlyuk, Pavel A. Ukrainskii, Zhanna A. Kirilenko, Olga A. Marinina, Olga M. Samofalova // Advances in Environmental Biology. 2014. Vol. 8. No 10: June. pp. 536-539.

Lisetskii F.N., Marinina O.A., Jakuschenko D.G. A new approach to dating the fallow lands in old-cultivated areas of the steppe zone // Research Journal of pharmaceutical, biological and chemical sciences. 2014. V. 5. No 6. P. 1325-1330.

Lisetskii F., Chepelev O. Quantitative substantiation of pedogenesis model key components // Advances in Environmental Biology. 2014. Vol. 8. Number 4: March. pp. 996-1000.

Lisetskii F. N., Goleusov P. V., Moysiyenko I. I., Sudnik-Wojcikowska B., Microzonal distribution of soils and plants along the catenas of mound structures // Contemporary Problems of Ecology. 2014. Vol. 7. No. 3. pp. 282–293.

Goleusov P. V., Lisetsky F. N. Restoration of soil and vegetation cover in post-mining geo-systems and their renaturation prospects in the area of the Kursk Magnetic Anomaly Gornyi Zhurnal. 2014. №8. C. 69–74.

Lisetskii F.N. Utilizzo come naturale antichi modelli di studio carriole pedo-geomorfologico rapporto. Italian Science Review. 2014; 6 (15). PP. 29-33. Available at URL: http://www.ias-journal.org/archive/2014/june/ Lisetskii.pdf.

Lisetskii F., Marinina O., Gadzhiev R. Trasformazione biomeccanica del profilo di suoli forestali. Italian Science Review. 2014; 12(21). PP. 134-137. Available at URL: http://www.ias-journal.org/archive/2014/ december/Lisetskii.pdf.

Lisetskii, F., Stolba, V.F., Marinina, O. Indicators of agricultural soil genesis under varying conditions of land use, Steppe Crimea (2015) Geoderma, 239-240, pp. 304-316. DOI: 10.1016/j.geoderma.2014.11.006.

Lisetskii F.N., Pichura V.I., Pavlyuk Y.V., Marinina O.A. Comparative assessment of methods for forecasting river runoff with different conditions of organization // Research Journal of Pharmaceutical, Biological and Chemical Sciences. 2015. Vol. 6. № 4. P. 56-60.

Yermolaev O.P., Lisetskii F.N., Marinina O.A., Buryak Zh.A. Basin and eco-regional approach to optimize the use of water and land resources // Biosciences, Biotechnology Research Asia, September 2015. Vol. 12(Spl. Edn. 2), p. 145-158.

Pichura V.I., Pilipenko Yu.V., Lisetskiy F.N., Dovbysh O.E. Forecasting of Hydrochemichal Regime of the Lower Dnieper Section using Neurotechnologies // Hydrobiological Journal. 2015. Vol. 51. No 3. pp. 100-110. DOI: 10.1615/HydrobJ.v51.i3.80.

Lisetskii F. N. and M. E. Rodionova. Transformation of dry-steppe soils under long-term agrogenic impacts in the area of ancient Olbia // Eurasian Soil Science. 2015. Vol. 48. No. 4. pp. 347–358. DOI: 10.1134/S1 064229315040055.

Scopus, Author ID: 15731091700
http://orcid.org/0000-0003-1273-5912

In: Terrestrial Biomes
Editor: Marlon Nguyen

ISBN: 978-1-63484-625-7


*Chapter 5*

# USE OF MICROORGANISMS AS AN ENVIRONMENTAL ALTERNATIVE TO TREAT AGRO-INDUSTRIAL WASTE

***Adriana M. Quinchía Figueroa[1,*], Juliana Uribe Castrillón[1], Verónica Cortés Durán[1] and Juan Carlos Loaiza Usuga[2]***

[1]Escuela de Ingeniería de Antioquia, Research Group on Sustainability, Infrastucture and Land SITE – A.A., Medellín, Colombia

[2]Universidad Nacional de Colombia - Sede Medellín, Facultad de Minas, Geosciences and Environmental Department, Colombia

## ABSTRACT

In recent years the Colombian productive system has been characterized by a low level of technology and a high use of inputs. This dynamic is associated with the increased use of pesticides in rural zones, which has harmful effects on the soil and water and has repercussions for human and animal health. This affects the wetland ecosystems located in high Andean mountains; as such, reducing the concentration of these substances is an environmental priority.

2,4,5,6-Tetrachloroisophthalonitrile is one of the chemical substances most commonly used as the active fungicidal ingredient in fungicides applied to fruit, vegetable and ornamental crops.

On the other hand, the solid waste derived from the fique (Furcraea andina) agroindustry is an environmental problem. The extraction of fique fiber from the plants generates solid waste that is thrown into water sources.

This research studied the fungicide biodegradation potential of activity by microorganisms isolated from fique solid waste.

* Corresponding author: Adriana Quinchía Figueroa. Escuela de Ingeniería de Antioquia; Research group on sustainability, infrastucture and land SITE – A.A. 7516 Medellín, Colombia. E-mail: adriana.quinchia@eia.edu.co.

The isolated strains were incubated in order to determine which ones were resistant to environments rich in Tetrachloroisophthalonitrile. The effect of these microorganisms on the biodegradation of this compound (19 g $L^{-1}$) was evaluated over a 3-day period, with a maximum of 99.85% degradation occurring.

The in vitro study showed that isolated and purified organisms from fique solid waste can grow in and tolerate concentrations of fungicide, efficiently degrading it. This indicates a high potential for biodegradation of other chlorate substances.

**Keywords:** biodegradation, pesticides, agro-industry residues, biotechnology, *Furcraea andina*

## INTRODUCTION

The fique plant (*Furcraea andina*) is a xerophytic monocot native to Colombia's Andean regions. Its production has expanded to Ecuador, Costa Rica, the Antilles, Brazil and Peru. Fibers extracted from the stiff leaves of the fique bush have been used to produce fique sacks, an industrial process that has existed for 45 years (Delgado et al., 2015). The fique plant is also known in some regions of Colombia as "cabuya" (Teles et al., 2015). Annual fique fiber production in Colombia exceeds 21,445 tons; just 4% of this corresponds to fiber, while 96% is solid waste (bagasse) or waste water (residual water) (Espinal, 2006; Bonilla, 2000; Castro, 2006).

Fique bagasse is a recalcitrant material that is difficult to degrade. Its effect centers on interference in the nutrient cycle of organisms (Barragán, 2007; Roldan, 2006; Rigas 2005; Caux, 1996).

In recent years, many studies have been conducted to document non-conventional medicinal systems, together with the use of their related pharmacopoeias (Gonzales de la Cruz et al., 2014).

Fique fiber (Delgado et al., 2015) has ethnobotanical applications (Hernández and León, 1994; Gonzales de la Cruz et al., 2014). By processing different parts of the plant, such as the leaves, stem, fruits, and even roots, it is possible to obtain an extract for engineering purposes, which can be used as a possible substitute for synthetic fiber in polymer composites (Monteiro et al., 2011; Teles et al., 2015).

The applications of natural fiber composites in construction are being investigated, due to the increasing demand for environmentally-friendly construction materials (Gómez and Vázquez, 2012), and to their effects on controlling soil erosion, retaining soil moisture, fixing atmospheric $CO_2$ and releasing $O_2$. Additionally, they can transform stored carbon into substances such as lignins, celluloses, hemicelluloses, proteins, sugars and solid waste or biomass, thus improving the physical and chemical characteristics of poor soils (Quinchia et al., 2013).

Nevertheless, inadequate management of these solid waste materials generates contamination, especially in water sources, due to the extremely acidic pH conditions that it creates. This is because these materials contain compounds such as saponines, hecogenines, and dodecanoic acid (Montoya, 1979; MAVDT- MADR-DNP, 2006).

Fique is normally planted in association with vegetable and fruit crops. Its production is characterized by a high use of inputs, including 2,4,5,6-Tetrachloroisophthalonitrile (a class II toxicity agrochemical). This product causes environmental problems associated with the contamination of soil and superficial and subterranean water sources (Espinal, 2006; Chaves, 2007; Ministerio de Agricultura, 2004).

Pesticides play an important role in enhancing agricultural productivity by reliably controlling harmful pests. However, many pesticides and their partially degraded products are highly persistent in the natural environment and cause undesirable effects on non-target lower and higher organisms (Thangadurai and Suresh, 2014).

There are significant differences in rhizosphere microbial carbon substrate utilization patterns and enzymatic activities among vegetation types (Yan et al., 2013). Nonetheless, the use of microorganisms to biodegrade pesticides by biological means has been found to be an environmentally and economically efficient method when compared to other physical and chemical methods (Abraham et al., 2014). For a detailed review of accelerated microbial degradation and increased degradation rates of pesticides in soils found in tropical environments, see Arbeli and Fuentes (2007).

Physical methods like heterogeneous photocatalysis have been used as treatment methods for removing toxic pollutants and converting them into innocuous end products such as $CO_2$, $H_2O$ and mineral acids (Sivagami et al., 2014). Some research has focused on simulating pesticide mass balance and degradation (Muñoz-Carpena et al., 2010, 2015). Vegetative filter strips (VFS) have been widely adopted as a way of limiting the transport of pesticides from adjacent fields through infiltration and reducing runoff flow volumes caused by pesticides coming into contact with vegetation and soil (Reichenberger et al., 2007; Muñoz-Carpena et al., 2015).

Other researchers have focused on studying the microorganisms used in the bioremediation of contaminated sites (Abraham et al., 2014; Thangadurai and Suresh, 2014) and the biodegradation of pollutants caused by microbial metabolic activity in mixed plant communities in riparian soils (Yang et al., 2013). The use of fresh residues in soil remediation has shown high potential for the remediation of biocides, possibly due to improved microbial diversity and enzymatic activity (Ribas et al., 2009). The use of biomixtures in combination with bioaugmentation strategies has been found to increase pesticide degradation efficiency (Karanasios et al., 2010; Pinto et al., 2016).

In this chapter, we studied the capacity of a combination of biomass and native microorganisms from fique bagasse from farms in Colombia's Eastern Antioquia region to degrade the active ingredient Tetrachloroisophthalonitrile (from Chlorothalonil fungicide).

This could make it possible to reduce the soil and water contamination risks associated with this pesticide, and could also create a use for agroindustrial solid waste. Additionally, the potential of this combination of biomass and native microorganisms to remediate pesticides will be evaluated.

## Materials and Methods

Fique bagasse was obtained from farms in San Vicente village, in the eastern part of Colombia's Antioquia department (Figure 1). An aliquot of solid waste was taken from the fiber extraction process. It was transported and refrigerated at 4ºC in hermetic bags.

The Tetrachloroisophthalonitrile used was ultrapure and a GC-μECD was used to test fungicidal reduction. The solvents used were all analytical (Panreac).

The microorganisms were isolated from fique bagasse. Fifty (50) ml of culture broth were added to two beakers in order to monitor bacterial and fungal growth; 5 gr of fique bagasse were added to each beaker and these were incubated for 24 hours at 37°C.

After this period, 10 ml of culture were taken from each beaker and were spread into nutritional agar and sabourand agar, respectively. The microbiological spread technique used was deep seed.

The agar petri dishes were incubated twice at 37°C, first for 24 hours to verify bacteria and yeast growth and then for a 7-day period to check fungus growth. (Each dilution and seeding had three repetitions.)

Microscopy analysis was used to identify individual strains, by evaluating size, color and type. The characteristics and distribution of the bacteria and yeast cultures were evaluated, and the spores and hyphae of the fungi were evaluated.

Once the active strains were isolated, Tetrachloroisophthalonitrile biodegradability testing was performed to evaluate the degradation capacity of these organisms. To prepare the substrate, the pesticide was applied to the medium until the concentration was 10 ppm. Afterwards, the medium was inoculated separately with the same amount of isolated fungi (*Rhizopus* sp and *Fusarium* sp) and yeast (*Candida boidinii).* It was then incubated at ambient temperature, with 100 rpm agitation in an aerated environment with an air filter.

From the start of the experiment and every 2 hours during the first 8 hours, the biomass ($mgL^{-1}$) and residual fungicide concentration were determined. After this period, these were measured every 24 hours for a 5-day period (Eweis, 1998).

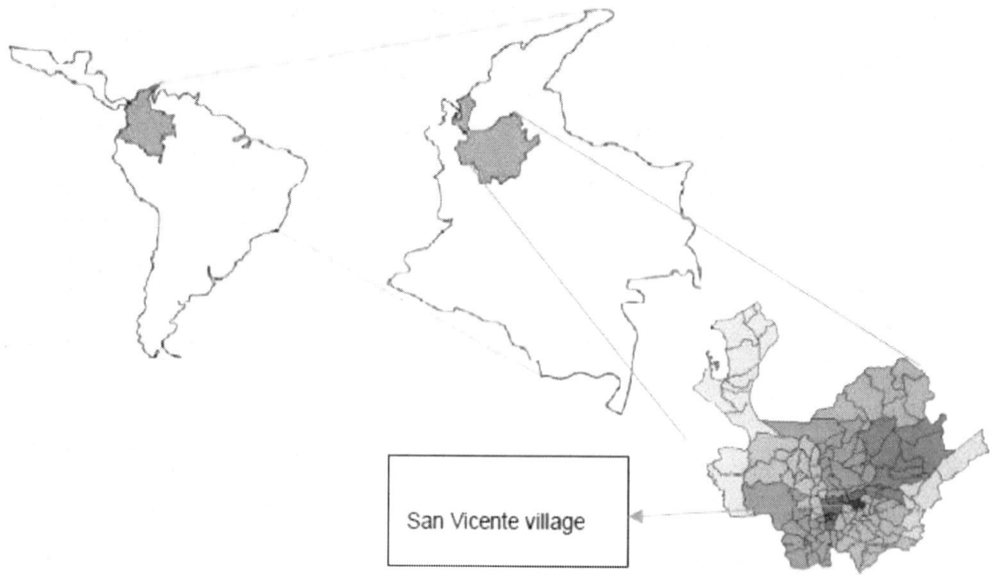


Figure 1. Study zone.

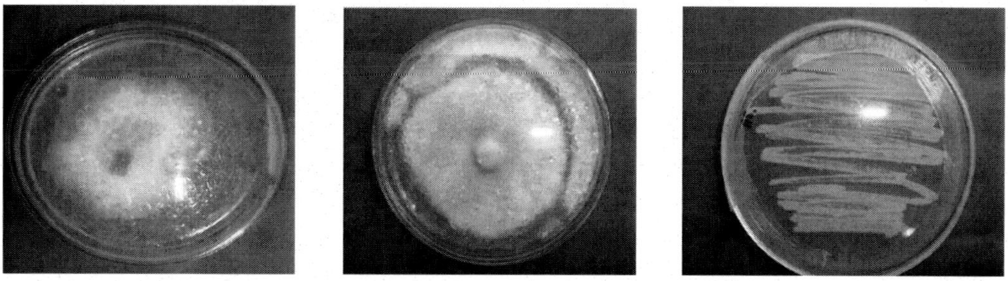

a b c

Rhizopus sp (a), Fusarium sp (b), Candida boidinii (c).

Figure 2. Strain isolated from fique bagasse.

To establish microorganism growth, three biomass proofs were performed for the strain that was growing. To quantify this evolution, 5 ml of the fermentation culture broth of each reactor were centrifuged at 6000 rpm for 10 minutes. Once the supernatant was eliminated, this was dried in an oven at 80ºC for 24 hours. The biomass was weighed and the dry biomass concentration per liter of solution (g $L^{-1}$ b.s.) was calculated.

# RESULTS AND DISCUSSION

## Identification and Microorganism Isolation

During the identification and isolation of microorganisms associated to fique bagasse with the capacity to grow in the presence of Tetrachloroisophthalonitrile, the Corporación para Investigaciones Biológicas (CIB) identified *Candida boidinii* yeast using the semi-automated assimilation method.

These microorganisms exhibited growth in 20 ppm of 2,4,5,6 Tetrachloro-isophthalonitrile. *Fusarium sp.* fungi have been reported to be capable of reducing toxic hexavalent chromium to its non-toxic trivalent form; degrading biodegradable plastics such as polybutylene succinate; and degrading chlorpyrifos, an organophosphate insecticide (Abe et al., 2010; Xie et al., 2010; Kumar et al., 2014). Meanwhile, Rhizopus sp. produce enhanced levels of free soluble phenolics as a potential nitrogen source; produce high antioxidant activity; and enhance waste decomposition (Correia et al., 2004). Some species have been used in the adsorption of copper, in the removal of thorium from solutions, and in the biosorption of americium and organochlorate pesticides (Rome and Gadd, 1987; Gadd and White, 1989; Liu et al., 2002; Ghosh et al., 2009).

*Candida boidinii* is important for regulating the synthesis of enzymes (enzymes that oxidize acidic n-amino acids) or metabolites. It is of great interest to biotechnology due to its action as an inhibitor (Fukunaga et al., 1998; Aggelis et al., 2000), and its great potential as a bioremediator and decontaminator.

## Growth Kinetics

To calculate the growth kinetics of the microorganisms studied, as well as Chlorothalonil (CLT) reduction, the fique microorganism isolates were tested in a liquid environment enriched with 10 ppm of Tetrachloroisophthalonitrile over the course of 5 days. Some microorganisms, such as *Serratia marcescens*, have achieved the complete biodegradation of organochloride pesticides such as DDT in concentrations up to 15 ppm. Similar studies have reported that the degradation process is inhibited at a 50 ppm concentration (Barragán, 2006; Chavez, 2003).

In this study, *Candida boidinii* presented exponential growth beginning at the time of inoculation, as can be seen in Figure 3.

There was no prolonged acclimation period observed in the stationary phase. This culture reached a maximum growth rate of 3.5 $gL^{-1}$, wich is higher than the results reported for other microorganisms, such as P. aeruginosa and F. oryzihabitants, which reached about 0.6 $gL^{-1}$ of biomass in the presence of organochlorate pesticides (Barragán et al. (2007).

Other studies with *C. boidinii* cultivated on methanol 33.6 $gL^{-1}$ at D_0.064 h_1 resulted in the production of 9.7 $gL^{-1}$ of dry biomass (Aggelis et al., 2000).

Biomass production decreased on the third day due to medium and nutrient depletion; however, great fungicide degradation was generated. These results were quite fast compared to endosulfan removal with *P. aeruginosa,* which reached only 51% in 7 days, according to Barragán et al. (2007).

To calculate the growth kinetics of the microorganisms studied, as well as Chlorothalonil (CLT) reduction, the fique microorganism isolates were tested in a liquid environment enriched with 10 ppm of Tetrachloroisophthalonitrile over the course of 5 days. Some microorganisms, such as *Serratia marcescens*, have achieved the complete biodegradation of organochloride pesticides such as DDT in concentrations up to 15 ppm. Similar studies have reported that the degradation process is inhibited at a 50 ppm concentration (Barragán, 2006; Chavez, 2003).

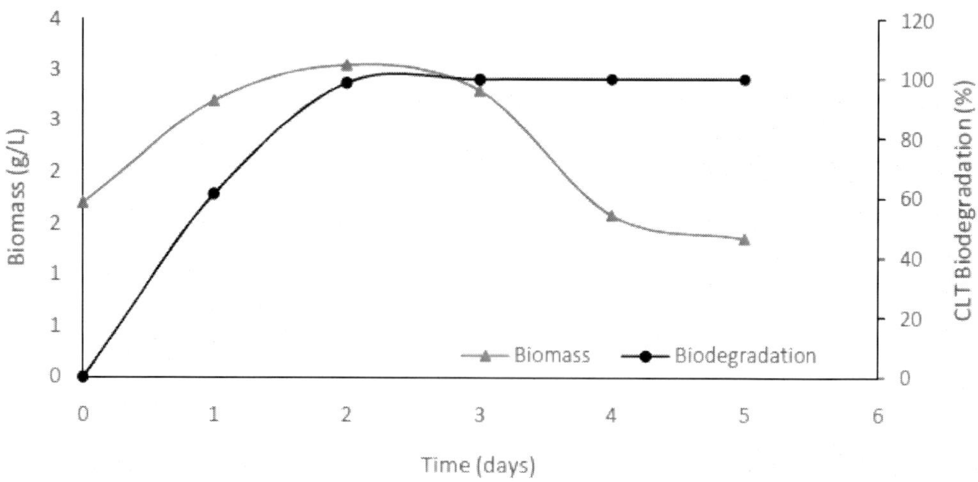


Figure 3. Growth kinetic and biodegradation percentage of CLT with *Candida boidinii.* Initial CLT concentration 20 ppm.

In this study, *Candida boidinii* presented exponential growth beginning at the time of inoculation, as can be seen in Figure 3.

There was no prolonged acclimation period observed in the stationary phase. This culture reached a maximum growth rate of 3.5 $gL^{-1}$, wich is higher than the results reported for other microorganisms, such as *P. aeruginosa* and *F. oryzihabitants*, which reached about 0.6 $gL^{-1}$ of biomass in the presence of organochlorate pesticides (Barragán et al. (2007).

Other studies with *C. boidinii* cultivated on methanol 33.6 $gL^{-1}$ at D_0.064 h_1 resulted in the production of 9.7 $gL^{-1}$ of dry biomass (Aggelis et al., 2000).

Biomass production decreased on the third day due to medium and nutrient depletion; however, great fungicide degradation was generated. These results were quite fast compared to endosulfan removal with *P. aeruginosa*, which reached only 51% in 7 days, according to Barragán et al. (2007).

*Rhizopus sp y Fusarium sp* fungi reached a maximum growing concentration of 18 g $L^{-1}$ (Figure 4) and 30 g $L^{-1}$ (Figure 5) respectively. With Rhizopus sp, it was observed that the microbiological culture had an adaptation phase that took an average of 24 hours (this coincides with research results for Aspergillus niger (Tejomyee et al., 2007)).

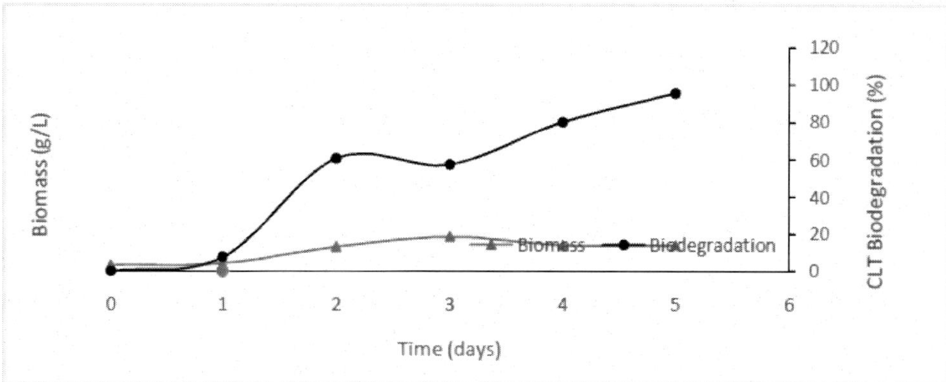


Figure 4. Growth kinetic and biodegradation percentage of CLT with *Rhizopus* sp. Initial CLT concentration 20 ppm.

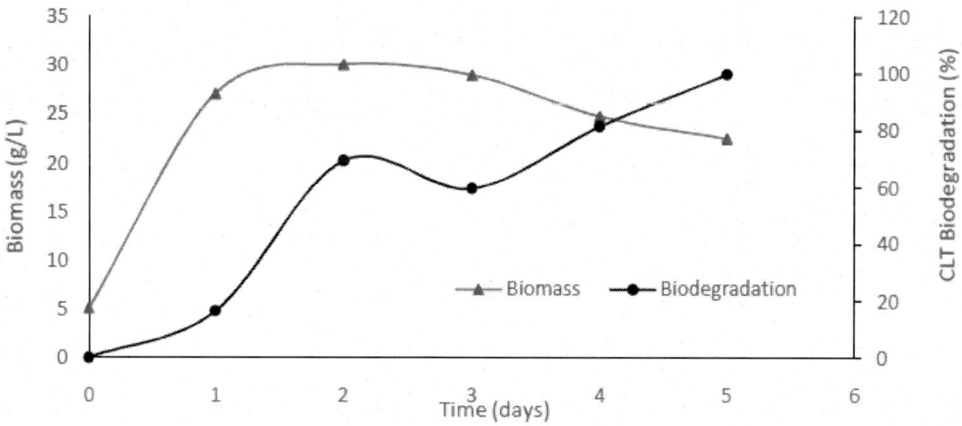


Figure 5. Growth kinetic and biodegradation percentage of CLT with *Fusarium* sp. Initial CLT concentration 20 ppm.

This adaptation period was not present in *Fusarium sp*, which presented cellular growth beginning at inoculation. In *Rhizopus sp*, accelerated growth of biomass production was observed; once this first phase had ended, cellular growth was derived from the degradation of the compound (Barragán, 2006). For *Rhizopus*, Correia et al. (2004) reported total phenolics decrease during the first 4 days of growth, with a subsequent increase until day 4.

In regard to biodegradation, both fungi cultures were found to reach the maximum biodegradation level on day 5. Yeast cultures are comparatively more efficient in regard to the time required to degrade the compound studied.

Based on the results obtained, the fungi studied here are more active than some other strains of fungi studied by Rigas et al. (2005). Other researchers (Tejomyee et al., 2007) found 2.6% lindane degradation and 181 mg of biomass production in 12.45 days using *Pl ostearatus*, at a lower pesticide concentration (2.03 ppm). Liu et al.(2002) reported that the *R. arrihizus* adsorption rate could be up to 99% in 1 hour when the pH was above 3, but decreased gradually as the pH value increased. Ghosh et al. (2009) reported that *R. oryzae* biomass was a good adsorbent for removing organochloride pesticides; the process was very fast, reaching equilibrium within 60 min.

The maximum biomass concentration of 19 g $L^{-1}$ was obtained with a joint culture comprised of *Candida boidinii, Rhizopus sp* and *Fusarium sp*. The behavior during microbial growth showed a short adaptation phase of less than 24 hours, followed by an exponential increase in biomass production (Figure 6). Studies carried out by Young and Banks (1998) and Ju et al. (1997), which investigated lindane adsorption by *R. oryzae* biomass and by certain gram positive and gram negative bacteria, respectively, reported higher lindane adsorption at lower pH.

This joint culture presented a CLT reduction higher than 99% over a period of up to 3 days, making it the best biodegredation option when compared with the other cultures studied. It is important to consider that in comparison, the compound removal obtained with the reference sample (pesticide degradation blank sample, without the presence of microorganisms) under similar conditions was 59%.

source.

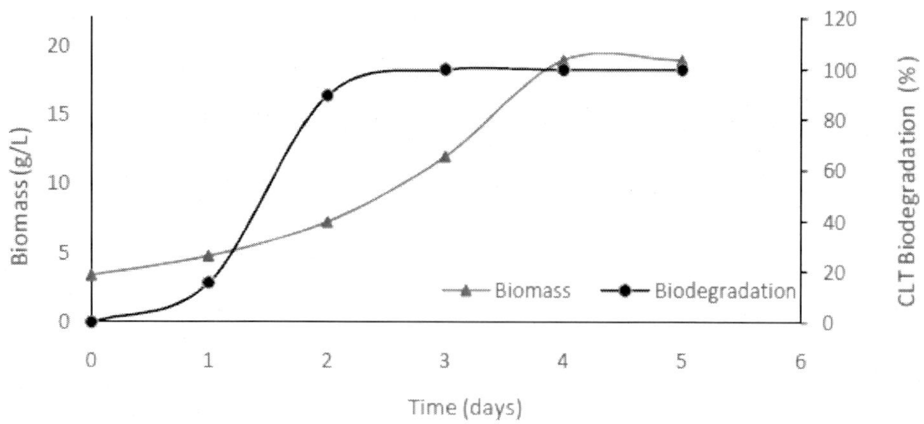


Figure 6. Growth kinetic of the joint culture (*Candida boidinii*, *Rhizopus* sp, *Fusarium* sp). Initial CLT concentration 20 ppm.

The most efficient biodegradation of the pesticide studied (also commercialized with the name Daconil) was achieved using yeast and the joint culture (99.85% and 99.7% over a period of 5 days and 3 days, respectively). These results are higher than those reported by Zhang et al. (2007), who achieved 91.3% biodegration of the same active compound using *Bacillus cereus* over a 6-day period in a culture enriched with glucose as a carbon

Barragán (2006), who studied the degradation of organochlorate pesticides such as DDT and endosulfan, achieved removal percentages of 51% and 30% using Pseudomona aeruginosa and coffee grains (as a carbon source) on glucose and peptone media, respectively, over a 7-day treatment period. Other authors found that 6% and 48.4% lindane biodegradation was achieved using Pleurotus ostreatus and Micrococcus varians, respectively, over a 12-15 day treatment period (Rigas et al., 2005; Abou-Arab, 2002).

In the literature there are other joint cultures (*Ochrobacterum sp., Arthrobacter sp.* and *Burkholderia sp.*) that have been found to biodegrade endosulfan (in contamined soils), with 72% and 87% degradation of the isomers α -endosulfan and β-endosulfan obtained after 20 days, respectively (Kazos et al., 2008). Biosorptive removal of the highly toxic organochlorate pesticide lindane from water has also been studied with both live and dead R. oryzae biomass of varying culture age. Depending on the age of the culture, lindane adsorption ranged from 63% to 90%. A decreasing trend was noted after the organism entered the death phase (Ghosh et al., 2009).

## CONCLUSION

Microorganisms isolated from fique can adapt to liquid environments. In future research, it is important to evaluate the effects of pH and temperature on this characteristic. In natural conditions, these microorganisms are found in solid environments associated to compounds specific to F. andina leaves. These include D-limonene, benzothiazole, phenol, ciclohexene-1-carboxaldehyde, 2,6,6-trimethyl, hecogenin acetate, spirostan-3-ol, stigmasterol, dodecanoic acid, pentadecanoic acid, n-hexadecanoic acid, eicosane, and tridecane (Álvarez, 2010; Martínez, 2002; Trejo, 2000).

Due to the molecular weight of these last compounds, the associated microorganisms have an advantage over other substances. The organisms studied were confirmed to be effective at degrading Tetrachlorois ophthalonitrile, the active ingredient of organochlorate pesticides, obtaining an efficiency of 99.85% over a 3-day period. Fique bagasse thus has clear environmental applications.

The behavior observed confirms the potential of *Rhizopus sp, Fusarium sp* and *Candida boidinii* as biosorbents that remove pesticides. This supports the information found in the literature regarding studies performed in laboratory conditions.

## ACKNOWLEDGMENT

We would like to give special thanks to the Escuela de Ingeniería de Antioquia for providing financial and technical support for this study.

## References

Abe, M., Kobayashi, K., Honma, N., Nakasaki, K. 2010. Microbial degradation of poly(butylene succinate) by Fusarium solani in soil environments. *Polymer Degradation and Stability*. 95. 138-143.

Abraham, J., Silambarasan, S., Logeswari, P. 2014. Simultaneous degradation of organophosphorus and organochlorine pesticides by bacterial consortium. *Journal of the Taiwan Institute of Chemical Engineers*. 45. 2590-2596.

Abou-Arab, A. 2002. Degradation of organochlorine pesticides by meat starter in liquid media and fermented sausage. *Food and Chemical Toxicology*. 33-41.

Aggelis, G., Margariti, N., Kralli, C., Flouri, F. 2000. Growth of Candida boidinii on methanol and the activity of methanol-degrading enzymes as affected from formaldehyde and methylformate. *Journal of Biotechnology*. 80. 119-125.

Arbeli, Z and Fuentes, C.L. 2007. Accelerated biodegradation of pesticides: An overview of the phenomenon, its basis and possible solutions; and a discussion on the tropical dimension (Accelerated biodegradation of pesticides: An overview of the phenomenon, its basis and possible solutions; and a discussion on the tropical dimension). *Crop Protection*. 26. 1733-1746.

Barragán, B. 2006. Aportación de los microorganismos asociados al grano verde de café en la degradación de plaguicidas organoclorados y su relación con la capacidad de adsorción del grano (Contribution of microorganisms associated with green coffee bean in the degradation of organochlorine pesticides and their relationship with the adsorption capacity of grain). Centro de Investigación y de Estudios Avanzados del Instituto Politécnico Nacional (CINVESTAV). México. (In Spanish).

Barragán, B., Costa, C., Peralta, J., Barrera, J., Esparza, F., and Rodriguez, R. 2007. Biodegradation of organochlorine pesticides by bacteria grown in microchines of the porous structure of green vean coffe. *International Biodeterioration and Biodegradation*. 239-244.

Bonilla, J., Peinado, J., Urdaneta, M., and Carrascal, E. 2000. Informe nacional sobre uso y manejo de plaguicidas en Colombia tendiente a identificar y proponer alternativas para reducir el escurrimiento de plaguicidas al mar Caribe (National report on the use and management of pesticides in Colombia aimed at identifying and proposing alternatives to reduce pesticide runoff into the Caribbean Sea). Ministerio del Medio Ambiente (Ministry of Environment), Colombia (In Spanish).

Castro, C and Palencia, A. 2006. Evaluación de la influencia de diferentes condiciones de tratamientos superficiales sobre el comportamiento de fibras de fique (Evaluating the influence of different conditions of surface treatments on behavior of sisal fibers). Facultad de Ingeniería Química (School of Chemical Engineering). Universidad Pontificia Bolivariana. Medellín, Colombia (In Spanish).

Caux Y. 1996. Environmental fate and effects of chlorothalonil: a Canadian perspective, *Critical Review Environmental Science Technology*. 45-93.

Chaves, A., Shea, D., Danehower, D. 2007. Analysis of chlorothalonil and degradation products in soil and water by GC/MS and LC/MS. *Chemosphere*. 629-638.

Chavez, B., Quintero, R., Esparza, F., Mesta, A.M., Zavala F.J., Hernández C.H., Gillen, T., Poggi H.M., Barrera, J., Rodriguez R. 2003. Removal of phenanthrene froam soil by co-cultures of bacteria and fungi pre-grown on sugarcane bagasse pith. Bio-*resource Technology*. 177-183.

Correia, R.T.P., McCue, P., Magalhães, M.M.A., Macedo, G. R., Shetty, K. 2004. Production of phenolic antioxidants by the solid - state bioconversion of pineapple waste mixed with soy flour using Rhizopus oligosporus. *Process Biochemistry*. 39. 2167-2172.

Delgado, F., Coronado, J. J., Rodríguez, S.A. 2015. Failure analysis of a machine support for fique fibre processing. *Engineering Failure Analysis*. 56. 58-68.

EPA. 2006. Environmental Protection Agency (EPA), US. June 2010. http://www.epa.gov.

Espinal, C., Martinez, H., Pinzón, N. 2006. La cadena del fique en Colombia: una mirada global a su estructura y dinámica (Fique chain in Colombia: a comprehensive look at its structure and dynamics). Ministerio de Agricultura y Desarrollo Rural (Ministry of Agriculture and Rural Development). Bogotá D.C. Colombia 21 p (In Spanish).

Eweis, J.B., Ergas, S.J., Chang, D., Schroeder, E. 1998. *Bioremediation principles*. Mc Graw Hill. Boston. 296 p.

Fukunaga, S., Yuno, S., Takahashi, M., Taguchi, S., kera, Y., Odani, S., yamada R. 1998. Purification and Properties of D-Glutamate Oxidase from Candida boidinii. *Journal of Fermentation and Bioengineering*. 85 (6). 579-583.

Ghosh, S., Da, S.K., Guha, A.K., Sanyal, A.K. 2009. Adsorption behavior of lindane on Rhizopus oryzae biomass: Physico-chemical studies. *Journal of Hazardous Materials*. 172. 485-490.

Gonzales de la Cruz, M., a, Baldeón, S., Beltrán, H., Jullian, V., Bourdy, G. 2014. Hot and cold: Medicinal plantuses in Quechua speaking communities in the high Andes (Callejón de Huaylas, Ancash, Perú). *Journal of Ethnopharmacology*. 155. 1093-1117.

Gómez, C and Vázquez, A. 2012. Flexural properties loss of unidirectional epoxy/fique composites immersed in water and alkaline medium for construction application. *Composites Part B: Engineering*. 43 (8). 3120-3130.

Hernández J.E. V and León, J. 1994.Neglected crops: 1492 from a different perspective. *FAO Plant Production and Protection Series, no. 26. 1. Title II. Series III.* FAO, Rome. Italy. 348 p.

Ju, Y., Chen, T., Liu, J.C. 1997. Study on the biosorption of lindane. *Colloids Surf. B: Biointerfaces*. 9. 187-196.

Karanasios, E., Tsiropoulos, N.G., Karpouzas, D.G., Menkissoglu-Spiroudi, U. 2010. Novel biomixtures based on local Mediterranean lignocellulosic materials: evaluation for use in biobed systems. *Chemosphere*. 80. 914-921.

Kazos E. A., Nanos C., Stalikas, C., Konidari C. 2008. Simultaneous determination of chlorothalonil and its metabolite 4-hydroxychlorothalonil in greenhouse. *Chemosphere*. 72. 1413-1419.

Kumar, M., Kumar, A., Bhattacharyya, M. 2014. A green chemical approach for biotransformation of Cr(VI) to Cr(III), utilizing Fusarium sp. MMT1 and consequent structural alteration of cell morphology. *Journal of Environmental Chemical Engineering*. 2. 424-433.

Liu, N., Yang, Y., Luo, S., Zhang, T., Jin, J., Liao, J., Hua, X. 2002. Biosorption of 241Am by Rhizopus arrihizus: preliminary investigation and evaluation. *Applied Radiation and Isotopes*. 57. 139-143.

Martínez, A and Caicedo, T. 2002. Bioensayo de toxicidad de los jugos de fique en peces en el municipio de Tambo (Toxicity bioassay sisal juice in fish in the town of Tambo) (Nariño)). Universidad El Bosque. Bogotá D.C. 26 p. (In Spanish).

MADR. 2004. Caracterización de la producción y la comercialización de plaguicidas en Colombia (Characterization of the production and marketing of pesticides in Colombia). Ministerio de Agricultura y Desarrollo Rural (MADR) (Ministry of Agriculture and Rural Development (MARD)). Bogota D.C. (In Spanish).

MAVDT., MADR., DNP. 2006. Guía ambiental del subsector fiquero. Ministerio de Ambiente, Vivienda y Desarrollo Territorial (MAVDT) (Environmental Guide fiquero subsector. Ministry of Environment, Housing and Territorial Development (MAVDT)), Ministerio de Agricultura y Desarrollo Rural (MADR) (Ministry of Agriculture and Rural Development (MARD)), Departamento Nacional de Planeación (DNP) (National Planning Department (DNP)). Editorial Panamericana Formas de Impresos S.A. Bogotá D.C. (Editorial Panamericana SA Printed Forms Bogotá D.C.) 121 p. (In Spanish).

Monteiro, S.N., Lopes, F.P.D., Barbosa, A.P., Bevitori, A.B., Silva, I.L.A., Costa, L.L. 2011. Natural lignocellulosic fibers as engineering materials - an overview. Metallurgical and Materials Transactions. *Physical Metallurgy and Materials Science*. 42 (10). 2963-2974.

Montoya, R and Tobón, H. 1979. Algunos aspectos sobre el cultivo de fique (*Furcraea spp*) y control de la calidad de la fibra (Some aspects of the cultivation of sisal (Furcraea spp) and control of fiber quality). Universidad Nacional de Colombia (National university of Colombia). Sede Medellin. Facultad de Agronomía. 63 p. (In Spanish).

Muñoz-Carpena, R., Fox, G.A., Sabbagh, G.J. 2010. Parameter importance and uncertainty in predicting runoff pesticide reduction with filter strips. *Journal Environmental Quality*. 39. 630-641.

Pinto, A.P., Rodrigues, S.C., Caldeira, A.T., Teixeira, D.M. 2016. Exploring the potential of novel biomixtures and Lentinula edodes fungus for the degradation of selected pesticides. Evaluation for use in biobed systems. *Science of the Total Environment*. 541. 1372-1381.

Quinchía, A., Uribe, J., Mesa, C. 2013. Effects of Incorporating of Furcraea Species Biomass in an Acidic Andisols. Chapter 10. In: *Plantations: Biodiversity, Carbon Sequestration, and Restoration*. Nova Science Publishers, Hauppauge, NY. 189-202.

Ribas, L.C.C., de Mendonça, M.M., Camelini, C.M., Soares C.H.L. 2009. Use of spent mushroom substrates from Agaricus subrufescens (syn. *A. blazei, A. brasiliensis*) and Lentinula edodes productions in the enrichment of a soil-based potting media for lettuce (Lactuca sativa) cultivation: Growth promotion and soil bioremediation. *Bioresource Technology*. 100. 4750-4757.

Reichenberger, S., Bach, M., Skitschak, A., Frede, H.-G., 2007. Mitigation strategies to reduce pesticide inputs into ground- and surface water and their effectiveness: a review. *Science of Total Environment*. 384, 1-35.

Rigas, F., Dritsa, V., Marchant, R., Papadopoulou, K., Avramides, E., Hatzianestis, I. 2005. Biodegradation on lindane by Pleurotus ostreatus via central composite design. *Environment International*. 191-196.

Roldan, G. 2006. Remoción de hidrocarburos en un sistema de biopilas con adición de texturizantes (Removal of hydrocarbons in a system with addition of texturizing biopiles). Cinvestav IPN. México. 35-101 (In Spanish).

Sivagami, K., Ravi Krishna, R., Swaminathan, Y. 2014. Photo catalytic degradation of pesticides in immobilized bead photo reactor under solar irradiation.

Tejomyee, B and Pravin, P. 2007. Biodegradation of organochlorine pesticide, endosulfan, by a fungal soil isolate, *Aspergillus niger. International Biodeterioration and Biodegradation.* 315-321.

Tele, M.C., Rodrigues, G., Amoy, P., Colorado, H., Muylaert, F., Neves, S. 2015. Fique Fiber Tensile Elastic Modulus Dependence with Diameter Using the Weibull Statistical Analysis. *Materials Research.* 7p.

Thangadurai, P and Suresh, S. 2014. Biodegradation of endosulfan by soil bacterial cultures. *International Biodeterioration and Biodegradation.* 94. 38-47.

Trejo M and Quintero R. 2000. Bioremediation of Contaminated Soils. *Environmental biotechnology and Cleaner Bioprocesses.* 179-190.

Xie, H., Zhu, L., Ma, T., Wang, J., Wang, J., Su, J., Shao, B. 2010. Immobilization of an enzyme from a Fusarium fungus WZ-I for chlorpyrifos degradation. *Journal of Environmental Sciences.* 22 (12). 1930-1935.

Yang, C., Yulai, W., Wang, M., Chen, H. 2013. Plant species influence microbial metabolic activity and butachlor biodegradation in a riparian soil from Chongming Island, China. *Geoderma.* 193, 194. 165-171.

Young, E and Banks, C.J. 1998. The removal of lindane from aqueous solution using a fungal biosorbent: the influence of pH, temperature, biomass concentration and culture age. *Environmental Technology.* 19. 619-625.

Zhang, Y., Lu, J., Wu, L., Chang, A., Frankenberger, W. 2007. Simultaneous removal of chlorothalonil and nitrate by *Bacilllus cereus* strain NS1. *Science of the Total Environment.* 383-387.

# INDEX

## D



## E



## F

## G



## H



## I

## J



## K



## L



## M

## N



## O



## P



## R

## S

## T



## U

## V



## W



## Y



## Z